BRIDGING GAPS IN ELEMENTARY MATHEMATICS

First Edition

Amanfi, C.
Bonna, O.

BRIDGING GAPS IN ELEMENTARY MATHEMATICS

AMANFI, CHRIS
BSc. (Hons.)

BONNA, OKYERE
MS.(Ed.), MBA

Library of Congress Number: Pending
ISBN: 978-1-61957-128-0
ISBN: 978-1-61957-129-7 (Electronic Version)
ISBN: 978-1-61957-150-1

Okab Publishing, LLC
P.O. Box 79029
Charlotte, NC 28271
okab.publishing@gmail.com

Cover Design and Graphics by: Mediasense

Disclaimer Statement

This study material is a studying resource and is not, in any way, a substitute for formal study material prescribed/recommended by the governing education body. There are no qualifications awarded for the use of this study material. Its contents do not constitute legal or other professional advice on any subject matter. The author does not accept any responsibility for any loss which may arise from reliance on information contained in this study material.

Contents

INTRODUCTION

We cannot get away from Mathematics in our daily lives; it is omnipresent. Whatever the path in life that we choose to follow, we are likely to encounter Mathematics, in some form.

This study supplement was created to improve upon the appreciation, and fundamental understanding of Mathematics, by employing the **"why"** of Mathematics approach. It was also created to ensure that those who did not have an opportunity in their earlier years to study mathematics, or those who did, but need better understanding, would be able to "**bridge**" the gaps formed in their education.

The material covers topics found in the curriculum at the Primary to the pre-Junior High School level. It is not intended to be a substitute for formally prescribed schooling study materials and text books, but it is a supplement that provides deeper understanding, real-life applications and different perspectives of the universal elements of the Mathematics curriculum. It effectively approaches the subject from the root, to help make it less abstract.

Specifically, it seeks to address the following:

- To provide deep explanation and practice of challenging topics found at the Primary level.
- To equip parents and teachers, to enable them to provide Mathematics support to their children and pupils, respectively.
- To serve as a tool for anybody who needs to establish a good foundation in Mathematics.
- To demonstrate the **"why"** of Mathematics, and make the subject more meaningful.
- To help overcome the phobias and fears of Mathematics for all.

Chapter
I

1.0   Numbers in various languages

Number	0	1	2	3	4	5	6	7	8
English	Zero	One	Two	Three	Four	Five	Six	Seven	Eight
Twi	Hwee	Baako	Mmienu	Mmiensa	Enan	Enum	Nsia	Nson	Nwotwe
Hausa	Sifili	Daya	Biyu	Uku	Hudu	Biyar	Shidda	Bakwai	Takwas
Yoruba	Odo	Okan	Meji	Meta	Merin	Marun	Mefa	Meje	Mejo
Ewe	Nadeke O	Deka	Eve	Eton	Ene	Aton	Aden	Adren	Enyi
Ga	Ekobɛ	Ekome	Enyɔ	Etɛ	Ejwe	Enumo	Ekpaa	Kpawo	Kpaanyɔ

Number	9	10	11	12	13	14
English	Nine	Ten	Eleven	Twelve	Thirteen	Fourteen
Twi	Nkron	Edu	Dubaako	Dumienu	Dumiensa	Dunan
Hausa	Tara	Goma	Goma sha daya	Goma sha biyu	Goma sha uku	Goma sha hudu
Yoruba	Mesan	Mewa	Mokanla	Mejila	Metala	Merinla
Ewe	Asheke	Ewo	Ewikeke	Ewieve	Ewieton	Ewiene
Ga	Nɛɛhu	Nyɔŋma	Nyɔŋma kɛ ekome	Nyɔŋma kɛ enyɔ	Nyɔŋma kɛ etɛ	Nyɔŋma kɛ ejwe

Number	15	16	17	18	19
English	Fifteen	Sixteen	Seventeen	Eighteen	Nineteen
Twi	Dunum	Dunsia	Dunson	Duwotwe	Dunkron
Hausa	Goma sha biyar	Goma sha shidda	Goma sha bakwai	Goma sha takwas	Goma sha tara
Yoruba	Medogun	Merindilogun	Metadilogun	Mejidilogun	Mokandilogun
Ewe	Ewiaton	Ewieade	Ewiadre	Ewienyi	Ewiasheke
Ga	Nyɔŋma kɛ enumo	Nyɔŋma kɛ ekpaa	Nyɔŋma kɛ kpawo	Nyɔŋma kɛ kpaanyɔ	Nyɔŋma kɛ nɛɛhu

Number	20	21	22	23	24
English	Twenty	Twenty-One	Twenty-Two	Twenty-Three	Twenty-Four
Twi	Aduonu	Aduonu Baako	Aduonu Mmienu	Aduonu Mmiensa	Aduonu Nan
Hausa	Ashirin	Ashirin sha daya	Ashirin sha biyu	Ashirin sha uku	Ashirin sha hudu
Yoruba	Ogun	Mokanlelogun	Mejilelogun	Metalelogun	Merinlelogun
Ewe	Blave	Blavevodeke	Blavevoeve	Blavevoeton	Blavevoene
Ga	Nyɔŋmai-enyɔ	Nyɔŋmai-enyɔ kɛ ekome	Nyɔŋmai-enyɔ kɛ enyɔ	Nyɔŋmai-enyɔ kɛ etɛ	nyɔŋmai-enyɔ kɛ ejwe

Number	25	26	27	28	29
English	Twenty-Five	Twenty-Six	Twenty-Seven	Twenty-Eight	Twenty-Nine
Twi	Aduonu Num	Aduonu Nsia	Aduonu Nson	Aduonu Nwotwi	Aduonu Nkron
Hausa	Ashirin sha biyar	Ashirin sha shidda	Ashirin sha bakwai	Ashirin sha takwas	Ashirin sha tara
Yoruba	Medogbon	Merindilogbon	Metadilogbon	Mejidilogbon	Mokandilogbon
Ewe	Blavevoaton	Blavevoade	Blavevoadre	Blavevoenyi	Blavevoasheke
Ga	Nyɔŋmai-enyɔ kɛ enumo	Nyɔŋmai-enyɔ kɛ ekpaa	Nyɔŋmai-enyɔ kɛ kpawo	Nyɔŋmai-enyɔ kɛ kpaanyɔ	Nyɔŋmai-enyɔ kɛ nɛɛhu

Number	30	31	32	33	34
English	Thirty	Thirty-One	Thirty-Two	Thirty-Three	Thirty-Four
Twi	Aduasa	Aduasa Baako	Aduasa Mmienu	Aduasa Mmiensa	Aduasa Nan
Hausa	Talatin	Talatin da daya	Talatin da biyu	Talatin da uku	Talatin da hudu
Yoruba	Ogbon	Mokanlelogbon	Mejilelogbon	Metalelogbon	Merinlelogbon
Ewe	Blato	Blatovodeke	Blatovoeve	Blatovoeton	Blatovoene
Ga	Nyɔŋmai-etɛ	Nyɔŋmai-etɛ kɛ ekome	Nyɔŋmai-etɛ kɛ enyɔ	Nyɔŋmai-etɛ kɛ etɛ	Nyɔŋmai-etɛ kɛ ejwe

Number	35	36	37	38
English	Thirty-Five	Thirty-Six	Thirty-Seven	Thirty-Eight
Twi	Aduasa Num	Aduasa Nsia	Aduasa Nson	Aduasa Nwotwi
Hausa	Talatin da biyar	Talatin da shida	Talatin da bakwai	Talatin da takwas
Yoruba	Marundilogoji	Merindilogoji	Metadilogoji	Mejidilogoji
Ewe	Blatovoaton	Blatovoade	Blatovoadre	Blatovoenyi
Ga	Nyɔŋmai-etɛ kɛ enumo	Nyɔŋmai-etɛ kɛ ekpaa	Nyɔŋmai-etɛ kɛ kpawo	Nyɔŋmai-etɛ kɛ kpaanyɔ

Number	39	40	41	42
English	Thirty-Nine	Forty	Forty-One	Forty-Two
Twi	Aduasa Nkron	Aduanan	Aduanan Baako	Aduanan Mmienu
Hausa	Talatin da tara	Arbain	Arbain da daya	Arbain da biyu
Yoruba	Mokandilogoji	Ogoji	Mokanlelogoji	Mejilelogoji
Ewe	Blatovoasheke	Blane	Blanevodeke	Blanevoeve
Ga	Nyɔŋmai-etɛ kɛ nɛɛhu	Nyɔŋmai-ejwɛ	Nyɔŋmai-ejwɛ kɛ ekome	Nyɔŋmai-ejwɛ kɛ enyɔ

Number	43	44	45	46
English	Forty-Three	Forty-Four	Forty-Five	Forty-Six
Twi	Aduanan Mmiensa	Aduanan Enan	Aduanan Enum	Aduanan Nsia
Hausa	Arbain da uku	Arbain da hudu	Arbain da biyar	Arbain da shidda
Yoruba	Metalelogoji	Merinlelogoji	Marundiladota	Merindiladota
Ewe	Blanevoeton	Blanevoene	Blanevoaton	Blanevoade
Ga	Nyɔŋmai-ejwɛ kɛ etɛ	Nyɔŋmai-ejwɛ kɛ ejwe	Nyɔŋmai-ejwɛ kɛ enumo	Nyɔŋmai-ejwɛ kɛ ekpaa

Number	47	48	49	50
English	Forty-Seven	Forty-Eight	Forty-Nine	Fifty
Twi	Aduanan Nson	Aduanan Nwotwi	Aduanan Nkron	Aduonum
Hausa	Arbain da bakwai	Arbain da takwas	Arbain da tara	Hamsin
Yoruba	Metadiladota	Mejidiladota	Mokandiladota	Adota
Ewe	Blanevoadre	Blanevoenyi	Blanevoasheke	Blaton
Ga	Nyɔŋmai-ejwɛ kɛ kpawo	Nyɔŋmai-ejwɛ kɛ kpaanyɔ	Nyɔŋmai-ejwɛ kɛ nɛɛhu	Nyɔŋmai- enumo

Number	60	70	80	90
English	Sixty	Seventy	Eighty	Ninety
Twi	Aduosia	Aduonson	Aduowotwe	Aduokron
Hausa	Sittin	Sabain	Tamanin	Tasani
Yoruba	Ogota	Adorin	Ogorin	Adorun
Ewe	Bladen	Blaadren	Blanyi	Blasheke
Ga	Nyɔŋmai-ekpaa	Nyɔŋmai-kpawo	Nyɔŋmai-kpaanyɔ	Nyɔŋmai-nɛɛhu

Number	100	200	300	400	500	600	700
English	Hundred	Two-Hundred	Three-Hundred	Four-Hundred	Five-Hundred	Six-Hundred	Seven-Hundred
Twi	Oha	Ahanu	Ahasa	Ahanan	Ahanum	Ahansia	Ahanson
Hausa	Dari	Dari biyu	Dari uku	Dari hudu	Dari biyar	Dari shidda	Dari bakwai
Yoruba	Ogorun	Ogorun Meji	Ogorun Meta	Ogorun Meri	Ogorun Marun	Ogorun Mefa	Ogorun Meje
Ewe	Alofa Deka	Alofa Eve	Alofa Eton	Alofa Ene	Alofa Aton	Alofa Ade	Alofa Adre
Ga	Oha	Ohai-enyɔ	Ohai-etɛ	Ohai-ejwe	Ohai-enumo	Ohai-ekpaa	Ohai-kpawo

Number	800	900	1000
English	Eight-Hundred	Nine-Hundred	Thousand
Twi	Ahanwotwe	Ahankron	Apem
Hausa	Dari takwas	Dari tari	Dubu
Yoruba	Ogorun Mejo	Ogorun Mesa	Egberun
Ewe	Alofa Enyi	Alofa Asheke	Akpe Deka
Ga	Ohai-kpaanyɔ	Ohai-nɛɛhu	Akpe

Table 1.1 – Translation of Arabic numerals into various dialects.

Chapter
2

2.0   Number line

- A number line is infinite; it can be extended to the left or to the right, infinitely.
- Numbers to the left of 0, are less than 0, and are called negative numbers. They become smaller in value, the further you go to the left from 0.
- Numbers to the right of 0, are greater than 0, and are called positive numbers. They become greater in value, the further you go to the right from 0.
- The number line is normally drawn in units of 1, but it can be constructed in units of less than 1 or more than 1, as shown in Figure 2.1.

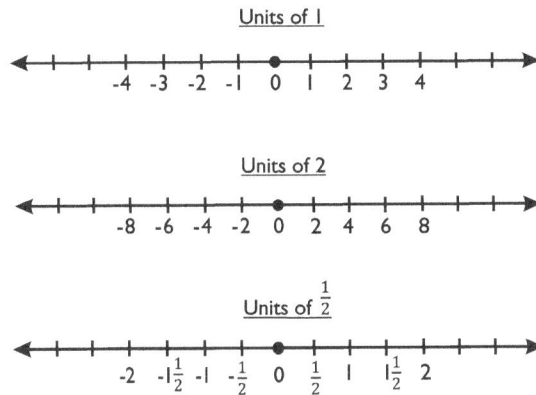

Units of 1

Units of 2

Units of $\frac{1}{2}$

Figure 2.1 – Number line in various units.

Application in Real Life

We use number lines in our lives all the time. For instance, we may need to use a tape measure to determine the length of a dress. A tape measure is designed on the basis of a number line.

Figure 2.2 – Tape measuring device.

Chapter

3 x 1

3.0 Counting numbers

3.1 How to count from zero in multiples

Method

A good way to approach this, is to draw a number line.

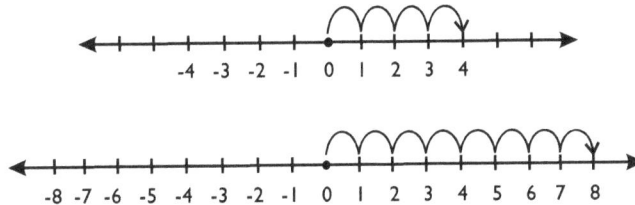

Figure 3.1 – Number lines in units of one.

For example, as shown in Figure 3.1, for multiples of 4, start from 0, and count four units to the right, to arrive at 4, then count further, another four units to the right, to arrive at 8, and so on. This tells us that $0 + 4 = 4$, $4 + 4 = 8$ and so forth.

Do the same for multiples of 8, by starting from 0, and counting eight units to the right, to arrive at 8, then count further, another eight units to the right, to arrive at 16, and so on. This also tells us that $0+8 = 8$, and $8+8 =16$ and so forth.

You can do this for any multiples that you wish.

Application in Real Life

We can use this to count quantities.

For example, how many dozens are in a crate of 24 eggs? The answer is 2 dozen, since a dozen is 12 and the crate has 24 eggs. So here, we are counting in units of 12.

Figure 3.2 – Crates of eggs.

Chapter 2^2

4.0   Counting backwards through zero to include negative numbers

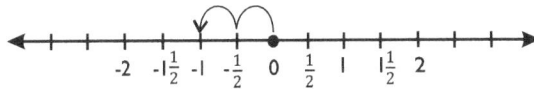

Figure 4.1 – Number line in units of $\frac{1}{2}$.

Any number less than zero is called a negative number. Remember, from Chapter 2, that numbers to the left of 0, on the number line, are less than 0 and are called negative numbers.

Here, we can also count in multiple units.

For example, for multiples of $\frac{1}{2}$, start from 0, and count $\frac{1}{2}$ units to the left, to arrive at $-\frac{1}{2}$, then count further, another $\frac{1}{2}$ units to the left, to arrive at -1, and so on. This tells us that:

$0+(-\frac{1}{2}) = -\frac{1}{2}$, and also, $-\frac{1}{2}+(-\frac{1}{2}) = -1$, and so forth.

Application in Real Life
If you have no money in your bank account, and you go overdrawn (go below zero balance), then you would owe money to the bank, since your bank account balance would now be a negative number. After seeing this negative account balance, your bank manager may then write a letter to you, asking you to restore your account to a zero balance.

Chapter

V

5.0 Filling in the blanks

Test your number skills by filling in the missing numbers in the grids below. The numbers in each grid should increase or decrease by the same quantity.

1	2							9	
		13			16				
21						27			30
				35					
				45					
51							58		
	62		64						
		83							
									100

100									91
	89								
		78							71
			67						
				56					
					45				
						34			
							23		
								12	
				6					1

0		10		20				40	
		60							

-30		-20		-10				10	
		30							

Chapter

5 + 1

6.0 Place value

This topic teaches us the value of a digit in a number.

We identify the place value of numbers, up to one billion, as follows (the dot is called a decimal point as you would see in Chapter 7).

Billion = 1,000,000,000.0 (the digit 1 is in the billions place).
Hundred Million = 100,000,000.0 (the digit 1 is in the hundred millions place).
Ten Million = 10,000,000.0 (the digit 1 is in the ten millions place).
Million = 1,000,000.0 (the digit 1 is in the millions place).
Hundred Thousand = 100,000.0 (the digit 1 is in the hundred thousands place).
Ten Thousand = 10,000.0 (the digit 1 is in the ten thousands place).
Thousand = 1,000.0 (the digit 1 is in the thousands place).
Hundred = 100.0 (the digit 1 is in the hundreds place).
Ten = 10.0 (the digit 1 is in the tens place).
Unit = 1.0 (the digit 1 is in the units place).
Tenth = 0.1 (the digit 1 is in the tenths place).
Hundredth = 0.01 (the digit 1 is in the hundredths place).
Thousandth = 0.001 (the digit 1 is in the thousandths place).
Ten Thousandth = 0.0001 (the digit 1 is in the ten thousandths place).
Hundred Thousandth = 0.00001 (the digit 1 is in the hundred thousandths place).
Millionth = 0.000001 (the digit 1 is in the millionths place).
Ten Millionth = 0.0000001 (the digit 1 is in the ten millionths place).
Hundred Millionth = 0.00000001 (the digit 1 is in the hundred millionths place).
Billionth = 0.000000001 (the digit 1 is in the billionths place).

Example

In the number 362:

Since 362 = 300 + 60 + 2, we have the following place values:

300 → Hundreds (that is, the digit 3 is in the hundreds place).
60 → Tens (that is, the digit 6 is in the tens place).
2 → Units (that is, the digit 2 is in the units place).

Application in Real Life
We learn place values to determine the value of a digit, based on its location in a given number.

Chapter
7

7.0   Decimals

The word Decimal simply means **"Based on 10"**.

Numbers to the left of a decimal point are whole numbers, and those to the right of the decimal point are parts of (less than) a whole number.

It is worthy to note that, any number can be written with a ".0" extension, where the dot is called a decimal point. For instance, the number 0 can be written as 0.0, the number 1 can be written as 1.0, the number 2 can be written as 2.0 and so on.

Examples

a. The number 2.0 has 2 units and 0 tenths, so it can be broken down as:

$$2.0 = (2 \times 1) + (0 \times \tfrac{1}{10}) = 2 + 0.0.$$

b. Here is the number "forty-five point six two", written as 45.62.

The number 45.62 has 4 tens, 5 units, 6 tenths and 2 hundredths. We break it down like this:

$$45.62 = (4 \times 10) + (5 \times 1) + (6 \times \tfrac{1}{10}) + (2 \times \tfrac{1}{100}) = 40 + 5 + \tfrac{6}{10} + \tfrac{2}{100} = 40 + 5 + 0.6 + 0.02$$

c. Here is the number "one thousand point zero six", written as 1000.06.

$$1000.06 = (1 \times 1000) + (0 \times 100) + (0 \times 10) + (0 \times 1) + (0 \times \tfrac{1}{10}) + (6 \times \tfrac{1}{100}) = 1000+0+0+0+0+.06$$

See Decimals on a number line as shown in Figure 7.1.

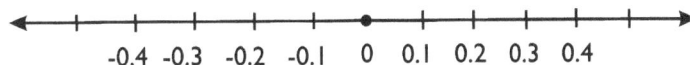

-0.4 -0.3 -0.2 -0.1   0   0.1   0.2 0.3 0.4

Figure 7.1 – Number line in units of 0.1.

Application in Real Life

- Decimals can be used, where precision is required. For example, if the price of a cup of coffee is 1 dollar and 23 cents, then this can be written as $1.23.
- Also in Athletics or Track and Field, times usually need to be recorded to decimal precision, to determine the finishing positions of the competitors. For instance, in a 100m race, 1st position = 9.86 seconds; 2nd position = 9.92 seconds and so on.

Chapter
7 + 1

8.0 Algebra

8.1 Addition

The Addition Table

+	1	2	3	4	5	6	7	8	9	10	11	12
1	2	3	4	5	6	7	8	9	10	11	12	13
2	3	4	5	6	7	8	9	10	11	12	13	14
3	4	5	6	7	8	9	10	11	12	13	14	15
4	5	6	7	8	9	10	11	12	13	14	15	16
5	6	7	8	9	10	11	12	13	14	15	16	17
6	7	8	9	10	11	12	13	14	15	16	17	18
7	8	9	10	11	12	13	14	15	16	17	18	19
8	9	10	11	12	13	14	15	16	17	18	19	20
9	10	11	12	13	14	15	16	17	18	19	20	21
10	11	12	13	14	15	16	17	18	19	20	21	22
11	12	13	14	15	16	17	18	19	20	21	22	23
12	13	14	15	16	17	18	19	20	21	22	23	24

Table 8.1 – Table showing addition of integers from 1 to 12.

In simple terms, addition means to add or to put together.

Method of Regrouping

Example

789 plus 642 is:

```
   789
+  642
  1431
   11
```

Following the steps below carefully, we begin solving this problem from the right to the left:

a) Starting from the 1st column from the right, we have $9 + 2 = 11$, so we regroup 11 into 1 "tens' and 1 "unit". We write 1, and then carry over 1 (see 1 carried over written below the 3 in the summation above) onto the next column on the left.

b) In the next column, we start by solving $8 + 4 = 12$. We then add on the 1 carried over from the first column to get 13. We cannot write 13, so we regroup 13 into 1 "tens" and 3 "units". We write 3, and then carry over 1 (see 1 carried over written below the 4 in the summation above) to the next column on the left.

c) In the third column from the right, we solve $7 + 6 = 13$. We then add on the 1 carried over from the previous column in b) to get 14. Here we can write 14, since there are no more columns left to carry over to, on the left.

Application in Real Life

If you have a car, and you buy an additional car, then you have added 1 car, to your existing car, to get 2 cars as the total number of cars you now have.

Figure 8.1.

8.2 Subtraction

Subtraction is the opposite of addition (just like to give is the opposite of to take).

Method of Regrouping

Examples

1. Solve 42 minus 27.

42 is made up of 4 "tens" and 2 "units". We attempt to do the subtraction, but since the 2 units is bigger than the 7 units, we regroup 42 into 3 "tens" and 12 "units" to enable the subtraction. So 12 – 7 = 5 and 3 - 2 = 1. Hence, the answer is 15.

2. 932 minus 457 is:

```
  932
- 457
  475
```

Here we start from the right and work our way back to the left as we did for the addition:
a) To solve 2 – 7, we regroup 32 into 2 "tens" and 12 "units". Hence 12 - 7 = 5.
b) Since we regrouped as in a), we are now left with 92 as opposed to 93. To solve 2 – 5, we regroup 92 into 8 "tens" and 12 "units". Hence 12 – 5 = 7.
c) We are now left with 8 in the hundreds column, after the regroup in part b).
d) Hence 8 – 4 = 4.

Application in Real Life

For instance, if you are given 50 dollars and then you spend 24 dollars out of it, you would need to subtract 24 from 50 to find out how much money you have left. It also helps you to determine if you have been given the right amount of change, after buying an item at a store.

Figure 8.2.

8.3 Multiplication

This is equivalent to a process of adding a number to itself a certain number of times. It is also known as the product.

The 12 times Table

X	1	2	3	4	5	6	7	8	9	10	11	12
1	1	2	3	4	5	6	7	8	9	10	11	12
2	2	4	6	8	10	12	14	16	18	20	22	24
3	3	6	9	12	15	18	21	24	27	30	33	36
4	4	8	12	16	20	24	28	32	36	40	44	48
5	5	10	15	20	25	30	35	40	45	50	55	60
6	6	12	18	24	30	36	42	48	54	60	66	72
7	7	14	21	28	35	42	49	56	63	70	77	84
8	8	16	24	32	40	48	56	64	72	80	88	96
9	9	18	27	36	45	54	63	72	81	90	99	108
10	10	20	30	40	50	60	70	80	90	100	110	120
11	11	22	33	44	55	66	77	88	99	110	121	132
12	12	24	36	48	60	72	84	96	108	120	132	144

Table 8.2 – Table showing multiplication of integers from 1 to 12.

8.3.1 Lattice Method of multiplication

This is how the lattice method works:

A lattice looks like a grid of columns and rows. Calculations are performed as follows:

If you are multiplying, for instance, a two-digit number by a two-digit number, your grid would have two columns and two rows. If you are multiplying a three-digit number by a three-digit number, your grid would have three columns and three rows. If you are multiplying a three-digit number by a two-digit number, your grid would have three columns and two rows (or two columns and three rows), and so on.

Then for each cell created in the grid, write a corresponding digit above and beside it. So, for 22 times 22, we would have:

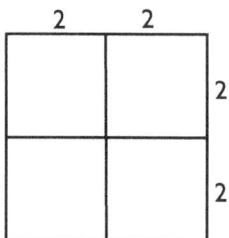

Then draw diagonal lines as below:

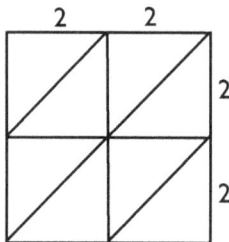

Next, multiply each number at the top by a number at the side and put the answer in the cell with the diagonal.

Then extend the diagonal lines, and add the numbers along the diagonals, to get 484 as the answer:

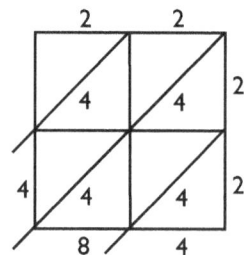

8.3.2 Short Multiplication

<u>Example</u>

2741 times 6 becomes:

```
  2741
x    6
16446
```

Here, starting from the right, we multiply each digit in the number above, 2741, by 6.

  a) $6 \times 1 = 6$
  b) $6 \times 4 = 24$; here we write 4 (the last digit in 24) and carry over the 2 to the next calculation.
  c) $6 \times 7 = 42$; here we add on the 2 from part b) above to get $42 + 2 = 44$. We write 4 and then carry over 4 to the next calculation.
  d) $6 \times 2 = 12$; add on the 4 from part c) above to get $12 + 4 = 16$.

8.3.3 Long Multiplication

<u>Example</u>

  1. 24 times 16 becomes:

```
   24
x  16
  144
+ 24
  384
```
    this line: (here, $6 \times 4 = 24$, so we write 4 and carry 2 over; $6 \times 2 = 12$, plus the carried over of 2 makes 14)
    this line: (here, $1 \times 4 = 4$ and $1 \times 2 = 2$; note where the 2 and 4 are positioned)

<u>Checking your answer using the Lattice Method</u>

24 times 16 is:

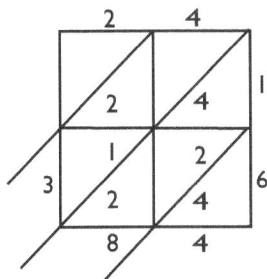

<u>Example</u>

   2.   35 times 55 becomes:

```
      35
  x   55
     175   this line: (here 5 x 5 = 25 so we write 5 and carry 2 over; 5 x 3 = 15, plus the carried over of 2 makes 17)
  + 175    this line: (here 5 x 5 = 25 so we write 5 and carry 2 over; 5 x 3 = 15, plus the carried over of 2 makes 17)
    1925
```

<u>Checking your answer using the Lattice Method</u>

35 times 55 is:

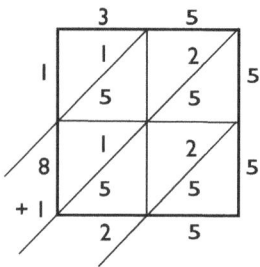

Application in Real Life

You can apply multiplication, when billing a customer. If you know the price of 1 item, and your customer buys 55 items from you, then you can use multiplication to get the total price of the 55 items.

8.4 Number Division

Division can be a challenge for children newer to Mathematics, and tends to be a turning point for them to conquer any fears they may have for Mathematics. Being able to master this topic goes a long way to being comfortable with Mathematics.

Division is the opposite of Multiplication.

Short division is similar to Long Division, except that, for Long Division, you would have more mathematical operations to perform.

8.4.1 Short Division

Because of its simplicity, pupils are normally introduced to division using this method.
We normally would use short division, when we want to divide a number that is up to 4 digits by a single digit number.

Example 1

98 divided by 7 becomes:

$$\begin{array}{r} 2 \\ 7\overline{\smash{\big)}98} \\ \hline 14 \end{array}$$

- a) 7 goes into 9 once (since 7 x 2 = 14, which is greater than 9 so we cannot use it), so we write 1. To find the remainder, we solve 9 – (7 x 1) = 2.
- b) We write the remainder of 2, in part a) above, as a superscript in the problem above to get 28.
- c) 7 goes into 28 four times, so we write 4.

So the answer is 14.

Example 2

49 divided by 11 becomes:

$$\begin{array}{r} 11\overline{\smash{\big)}49} \\ \hline 4 \ ^{r5} \end{array}$$

- a) 11 goes into 4 zero times, because 11 is bigger than 4, so we move on to the next solvable item.
- b) 11 goes into 49 four times, so we write 4 with a remainder of 5 (since 11 x 4 is equal to 44, and 44 is 5 less than 49).

So the answer is $4 + \frac{5}{11}$.

8.4.2 Long Division

We normally use Long Division, when we are dividing a number that has 3 or more digits by a number that has 2 or more digits.

<u>Example</u>

432 divided by 15 becomes:

$$
\begin{array}{r}
28\ {}^{r12} \\
15\overline{\smash{)}432} \\
\underline{30} \\
132 \\
\underline{120} \\
12 \\
\end{array}
$$

a) 15 goes into 4 zero times, since 15 is bigger than 4.
b) 15 goes into 43 two times, with a remainder of 13 (since 15 x 2 = 30 and hence, 43 minus 30 gives us 13).
c) 15 goes into 132 eight times, with a remainder of 12 (since 15 x 8 = 120 and hence, 132 minus 120 gives us 12).

So the answer is $28 + \frac{12}{15}$.

<u>Application in Real Life</u>

When you go out with 4 of your friends to a restaurant, and you order one large pizza, you may need to know how many parts to divide the pizza into, so everyone has a piece. Since you and your 4 friends make up 5 people in total, everyone's share of the pizza would equal $\frac{1}{5}$ (one divided by five).

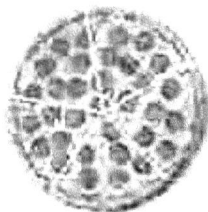

Figure 8.3.

Chapter 9

9.0 Fractions

In a fraction, the denominator (the lower part) tells us how many parts the numerator (the upper part) is being divided into.

Example

$\frac{3}{4}$ means 3 is divided into 4 parts. Here 3 is called the **numerator,** and 4 is called the **denominator.**

9.1 Types of fraction

Proper Fraction: In this type of fraction, the numerator is less than the denominator.

Examples: $\frac{3}{4}, \frac{2}{3}, \frac{4}{9}, \frac{93}{100}$.

Improper Fraction: In this type of fraction, the numerator is greater than or equal to the denominator.

Examples: $\frac{4}{3}, \frac{19}{7}, \frac{7}{2}, \frac{8}{8}$.

Mixed Fraction: An expression that consists of a whole number and a proper fraction.

Examples: $3\frac{1}{3}, 9\frac{3}{7}, 1\frac{5}{8}$.

We can convert a mixed fraction into an improper fraction .

In general:

$$a\frac{b}{c} = \frac{((a \times c)+b)}{c}.$$

Examples

a) $3\frac{1}{3} = \frac{((3 \times 3)+1)}{3} = \frac{10}{3}$.

b) $9\frac{3}{7} = \frac{((7 \times 9)+3)}{7} = \frac{66}{7}$.

c) $1\frac{5}{8} = \frac{((8 \times 1)+5)}{8} = \frac{13}{8}$.

Equivalent Fraction: These are fractions that are equal in value.

<u>Example</u>

$\frac{4}{8} = \frac{2}{4}$. Here $\frac{4}{8}$ reduces to $\frac{2}{4}$, when we divide both the 4 and the 8, in the fraction $\frac{4}{8}$, by the same number, 2.

Reciprocal: Reciprocals occur when the product of two fractions is equal to 1. You may recall that product simply means multiplication.

Generally:

The reciprocal of $\frac{a}{b}$ is $\frac{b}{a}$, since $\frac{a}{b} \times \frac{b}{a} = 1$.

<u>Example</u>

The reciprocal of $\frac{3}{4}$ is $\frac{4}{3}$, since $\frac{3}{4} \times \frac{4}{3} = 1$.

9.2 Addition and Subtraction of fractions with same denominator

In general:

a) $\frac{a}{b} + \frac{c}{b} = \frac{(a+c)}{b}$.

b) $\frac{a}{b} - \frac{c}{b} = \frac{(a-c)}{b}$.

<u>Examples</u>

i) $\frac{5}{7} + \frac{1}{7} = \frac{(5+1)}{7} = \frac{6}{7}$.

ii) $\frac{5}{7} - \frac{1}{7} = \frac{(5-1)}{7} = \frac{4}{7}$.

9.3 Ordering unit fractions with same denominator

<u>Example</u>

Here, given $\frac{6}{7}$ and $\frac{5}{7}$, we attempt to find out which fraction is greater.

<u>Answer</u>

Since the denominators are the same, in this case 7, we only compare the numerators. 6 is bigger than 5, therefore the fraction $\frac{6}{7}$, is greater than the fraction $\frac{5}{7}$.

9.4 Ordering unit fractions with different denominators

Example

Here, given $\frac{1}{10}$ and $\frac{2}{5}$, we attempt to find out which fraction is greater.

Answer

A simple way to do this, is to, first, convert the numbers to decimals and then perform the comparison on a number line.

$\frac{1}{10} = 0.1$, using short or long division.

$\frac{2}{5} = 0.4$, using short or long division.

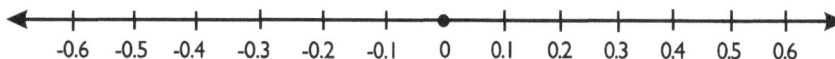

Figure 9.1 – Number line in units of 0.1.

So by inspection of Figure 9.1, we see that $\frac{2}{5}$, which is equal to 0.4, is greater than $\frac{1}{10}$, which is equal to 0.1.

Chapter
X

10.0 Ratios and Proportions

10.1 Ratios

For two numbers, a and b, their ratio is written as a:b; which simply indicates how many times the number b contains the number a.

For instance, if you own two houses, in a community of five houses, then the ratio of the houses you own, to the total number of houses in the community, is written as 2:5.

Ratios are used in many other situations.

Another example is your score in a test. If a Mathematics test has 40 questions, and you answered 35 of the questions correctly, then the ratio of your score to the total score is written as 35:40.

10.2 Equivalence of Ratios

Equivalent ratios are equal in value. For instance, the ratios, 1:3 and 3:9, are equivalent. To demonstrate this, we convert the ratios into fractions:

$1{:}3 = \frac{1}{3}$ and

$3{:}9 = \frac{3}{9} = \frac{1}{3}$.

So we see that they are equal in value.

10.3 Proportions

By setting up the ratios 1:3 and 3:9, if we write them as fractions $\frac{1}{3}$ and $\frac{3}{9}$ respectively, we see that they are equal in value.

By writing $\frac{1}{3} = \frac{3}{9}$, we have written a proportion. So we say here that " $\frac{1}{3}$ is proportional to $\frac{3}{9}$ ".

In real life, if we have a known ratio, when comparing two items, then we can predict the second ratio if we have information on one number of the second ratio.

Example

We are given 1 slice of cake, from a cake made up of 3 slices in total. There is a second cake, made up of 9 slices in total. If we are to receive the same proportion of slices from the second cake, as we did from the first cake, how many slices should we receive from the second cake?

Answer

Here, the ratio of slices to the total number of slices received from the first cake = 1:3.

Also, if we are to receive x slices for the second cake, then the ratio of slices, to the total number of slices, to be received in the second cake = x:9.

So to check for equivalence:

We write 1:3 as a fraction, $\frac{1}{3}$, and we write x:9 as a fraction, $\frac{x}{9}$.

Therefore: $\frac{1}{3} = \frac{x}{9}$, and x = 3 slices.

Chapter
XI

11.0 Manipulating Functions & Equations

A Function is a mathematical expression, and an equation can be described as a process or statement showing that two mathematical expressions are equal.

<u>Example 1</u>

If $3x - 5 = 7$, find x.

Here, in words, we are saying that 3 times a number called x, minus 5, is equal to 7.

Since the expressions on each side of the equal sign are equal in value, if we add, divide or subtract the same number to both sides, the expressions would still be equal to each other.

So the first thing we do, <u>in trying to isolate x</u>, is to:

Add 5 to both sides to get:

$3x - 5 + 5 = 7 + 5.$

The reason we choose to add 5, and not any other number to both sides, is because adding it would eliminate the -5 in the expression $3x - 5$ and reduce it to 3x only (note that 3x means 3 times x).

This simplifies to:

$3x = 12.$

Then we divide both sides by 3 to get:

$$\frac{3x}{3} = \frac{12}{3}.$$

The reason we choose to divide by 3, and not any other number, is because dividing by 3 would reduce the expression 3x to x only. It is good for us since we are trying to find the value of x only.

This simplifies to $x = 4$.

Therefore, our answer for x is 4.

Example 2

If $2x - 5 = 7 - x$, find x.

Here in words, we are saying that:

2 times a number called x, minus 5 is equal to 7 minus the same number called x.

As in Example 1:

In trying to isolate x, we will add x to both sides to get:

$2x + x - 5 = 7 - x + x$.

The reason we choose to add x to both sides, is to enable us to simplify the expression $7 - x$, to 7 only.

This simplifies to $3x - 5 = 7$.

Next, add 5 to both sides to get:

$3x - 5 + 5 = 7 + 5$.

The reason we choose to add 5, and not any other number to both sides, is because adding 5 would eliminate the -5 in the expression $3x - 5$ and reduce it to 3x only.

This simplifies to $3x = 12$.

Then we divide both sides by 3 to get:

$$\frac{3x}{3} = \frac{12}{3}.$$

The reason we choose to divide by 3, and not any other number, is because dividing by 3 would reduce the expression 3x to x only. It is good for us since we are trying to find the value of x only.

This simplifies to $x = 4$.

Therefore, our answer for x is 4.

Example 3

If $2x - y = 7 - x$, find x.

Here in words, we are saying that:

2 times a number called x, minus an unknown number called y is equal to 7 minus the same number called x.

We are trying to find x so our goal is to isolate x in the function.

We add x to both sides to get:

$2x + x - y = 7 - x + x.$

This simplifies to $3x - y = 7$.

Then we do the following:

Add y to both sides to get:

$3x - y + y = 7 + y.$

The reason we choose to add y, and not any other number to both sides, is because adding it would eliminate the -y in the expression $3x - y$, and reduce it to 3x only.

This simplifies to:

$3x = 7 + y.$

Then we divide both sides by 3 to get:

$$\frac{3x}{3} = \frac{(7+y)}{3}.$$

The reason we choose to divide by 3, and not any other number, is because dividing by 3 would reduce the expression 3x to x only. It is good for us since we are trying to find the value of x only.

This simplifies to:

$x = \frac{(7+y)}{3}$, which is our answer.

Our answer, this time, is not an absolute number, because the function we are trying to solve has 2 unknown variables x and y.

Application in Real Life

If doughnuts are sold in packs of 4, and we need 100 doughnuts for a party and 20 to give away to the needy, how many packs should we order to get 100 doughnuts for the party?

Here we set up the equation $4x - 20 = 100$, where x is the number of packs we need to order, which is 30 packs in this situation.

Chapter 12

12.0 Inequalities

We think of inequalities as any potential situation, aside of absolute equality.

For something not to be equal, then it can be:

1. <u>Less than (denoted <)</u>

    <u>Example</u>
    If $7x - 5 < 3$, then what is x?

    Here we solve for x in a similar way that we did for the examples in the equations:

    We add 5 to both sides to get:
    $7x - 5 + 5 < 3 + 5$.

    Which simplifies to:
    $7x < 8$.

    And we divide both sides by 7 to get, $x < \frac{8}{7}$, as the answer.

    We then check this answer by replacing x in the original equation with $\frac{8}{7}$, to get:

    $7(\frac{8}{7}) - 5 < 3$.

    This gives us the result $3 < 3$, which is not possible, so x has to be less than $\frac{8}{7}$. Hence our result is right.

2. <u>Greater than (denoted >)</u>

    <u>Example</u>
    If $7x - 5 > 3$, then what is x?

    Here we solve for x in a similar way that we did for the examples in the equations:

    We add 5 to both sides to get:
    $7x - 5 + 5 > 3 + 5$.

    Which simplifies to:
    $7x > 8$.

    And we divide both sides by 7 to get, $x > \frac{8}{7}$, as the answer.

    We then check this answer by replacing x in the original equation with $\frac{8}{7}$ to get:

    $7(\frac{8}{7}) - 5 > 3$.

This gives us the result $3 > 3$, which is not possible. so x has to be greater than $\frac{8}{7}$. Hence our result is right.

3. Less than or equal to (denoted ≤)

    Example
    If $7x - 5 \leq 3$, then what is x?

    Here we solve for x in a similar way that we did for the examples in the equations:

    We add 5 to both sides to get:
    $7x - 5 + 5 \leq 3 + 5$.

    Which simplifies to:
    $7x \leq 8$

    And we divide both sides by 7, to get, $x \leq \frac{8}{7}$, as the answer.

    We then check this answer by replacing x in the original equation with $\frac{8}{7}$ to get:
    $7(\frac{8}{7}) - 5 \leq 3$.

    This gives us the result $3 \leq 3$, which is valid, hence our result is right.

4. Greater than or equal to (≥)

    Example
    If $7x - 5 \geq 3$, then what is x?

    Here, we solve for x in a similar way that we did for the examples in the equations:

    We add 5 to both sides to get:
    $7x - 5 + 5 \geq 3 + 5$.

    Which simplifies to:
    $7x \geq 8$.

    And we divide both sides by 7 to get, $x \geq \frac{8}{7}$, as the answer.

    We then check this answer by replacing x in the original equation with $\frac{8}{7}$ to get:

    $7(\frac{8}{7}) - 5 \geq 3$.
    This gives us the result $3 \geq 3$, which is valid, hence our result is right.

Application in Real Life

Inequalities are used in so many ways in real life. For example:

1. You may need a minimum number to pass a Mathematics test (we write: $X \geq 60\%$, where 60% is the pass mark).

2. You may need to make a minimum monthly payment on a credit card balance (we write: $X \geq$ 10% of the credit card balance, where 10% of the credit card balance is the minimum payment to be made).

3. You may be allowed a maximum number of free texts on your cell phone per month (we write: $X \leq 5$, where 5 is the maximum number of free texts per month).

Chapter
12 + 1

13.0   Commutative Property

To commute simply means to move or to change places. This property works for addition and multiplication only.

If a = apples and b = oranges, then:

apples + oranges = oranges + apples.

Here we are saying that a + b = b + a.

or

If c =1 and d = 2,

Then, 1 x 2 = 2 x 1.

Here, we are saying that c x d = d x c.

Examples:

    a) 5 + 10 = 10 + 5.

    b) X + Y = Y + X.

    c) 5 x 10 = 10 x 5.

    d) X x Y = Y x X.

Application in Real Life

This tells us that the order in doing things are sometimes different, but can give us the same result. Example you can first put 5 oranges in a basket and then add 5 apples afterwards, or you can first put 5 apples in the basket and then add 5 oranges afterwards. Both would get you the same number and type of fruits in the basket.

Figure 13.1.

Chapter 14

14.0 Associative Property

This simply means to create associations or groupings. This property works for addition and multiplication only.

Here we are saying that given numbers a, b and c:

i.   (a + b) + c = a + (b + c) or
ii.   (a x b) x c = a x (b x c).

Examples:

   a) (2 + 3) + 4 = 2 + (3 + 4).
   b) (X + Y) + Z = X + (Y + Z).
   c) (2 x 3) x 4 = 2 x (3 x 4).
   d) (X x Y) x Z = X x (Y x Z).

Application in Real Life

For instance, you are given some money as a present, and then you decide to go shopping. You decide to buy a pair of shoes at 30 dollars, a pair of socks at 10 dollars and a handkerchief at 2.67 dollars. To make it easier to calculate how much you have spent, you can decide to add the 30 dollars to the 10 dollars first, and then add the total to the 2.67 dollars.

Chapter
XV

15.0 Distributive Property

Here we are saying that, given numbers a, b and c:

a (b + c) = a x (b + c) = (a x b) + (a x c).

Examples:

a) 2 (3 + 4) = 2 x (3 + 4) = (2 x 3) + (2 x 4).
b) 3 (4 – 2X + 5Y) = 3 x (4 – 2X + 5Y) = (3 x 4) – (3 x 2X) + (3 x 5Y).

Application in Real Life

You are in a class of 40 pupils and your teacher wants to buy each pupil 2 pencils and 3 erasers (A pencil and an eraser are all called instruments). How many instruments in total does he need to buy? Rather than counting one after the other, he can simply use the distributive property by solving 40(2 + 3) = 40 x (2 + 3) = 200 instruments.

Chapter 16

16.0 Measurement

We perform measurements using the Imperial or Metric system, though the Imperial system is not often used in Mathematics. Below are some key units of measurement.

16.1 Length & Width

Length measures the longest side of an object, and width measures how wide the object is.

Metric

1,000 Meters (m) = 1 Kilometer (km).
1,000 Millimeters (mm) = 1 Meter (m).
100 Centimeters (cm) = 1 Meter (m).
10 Decimeters (dm) = 1 Meter (m).

Imperial
12 Inches = 1 Foot.
3 Feet = 1 Yard.
1,760 Yards = 1 Mile.

Application in Real Life

We may need to find the Length of a piece of land, in order to be able to price it. This is done in standard units, such as the meter.

Figure 16.1.

16.2 Mass

Metric
1,000 Grams (g) = 1 Kilogram (kg).

Imperial
1 Pound = 16 Ounces.

Application in Real Life

We may need to find the Mass of luggage (how heavy our bags are), when loading luggage onto an airplane to ensure that the airplane can take off safely. A pilot does the overall calculation before he or she takes off into the air. This is usually done in kilograms or pounds.

Figure 16.2.

16.3 Volume

Metric

1,000 Millilitres (ml)= 1 Litre (l).
1,000 Litres (l) = 1 Cubic Meter (m³).
1 Cubic Centimeter (cm³) = 0.001 Litre (l).
1,000,000 Cubic Centimeters (cm³) = 1 Cubic Meter (m³).

Imperial

1 Pint = 20 Fluid Ounces.
1 Quart = 2 Pints.
1 Gallon = 4 Quarts.

Application in Real Life

We may need to find the volume of a barrel in order to determine how much water or oil it can carry.

Figure 16.3.

16.4 Time

Below are some widely used measures of time:

1 Minute = 60 Seconds.
1 Hour = 60 Minutes.
1 Day = 24 Hours.

Application in Real Life

Watches (seconds and hours) and calendars (days) are based on these measures of time.

Figure 16.4.

16.5 Temperature

Temperature measures the degree of hotness of an object.

The basic units of measurement are in degrees Celsius or degrees Fahrenheit.

Degrees Celsius (°C) = (Degrees Fahrenheit (°F) – 32) x $\frac{5}{9}$.

Application in Real Life

Thermometers, used to measure our body temperature, are in units of Celsius or Fahrenheit.

Figure 16.5.

Chapter
$$4^2 + 1$$

17.0 Geometry

17.1 Angles

The space between two lines, at the point they intersect, is called an angle.

17.1.1 Types of Angles

- Right-angle – This is an angle which measures exactly ninety degrees (90°).
- Acute angle – An acute angle is an angle less than ninety degrees (90°) in measurement.
- Straight angle – A straight angle measures exactly one-hundred and eighty degrees (180°).
- Obtuse angle – An obtuse angle is an angle more than ninety degrees (90°) and less than one-hundred and eighty degrees in measurement (180°).
- Reflex angle – This is an angle more than one-hundred and eighty degrees (180°) but less than three hundred and sixty degrees (360°).

17.1.2 Properties of Angles

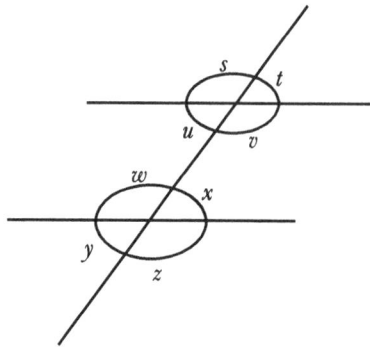

Figure 17.1.

Below are some angle properties of parallel lines. As per Figure 17.1:

- Corresponding angles are always equal. Therefore, $s = w$.
- Interior angles always add up to 180°. So, $u + w = 180°$.
- Alternate angles are equal. Hence, $u = x$.
- Vertically opposite angles are equal. So we have $s = v$.
- Adjacent angles add up to 180° and are supplementary. Therefore $s + t = 180°$.

17.2 Basic Trigonometry

A right-angle triangle is a triangle in which one of its angles measures ninety degrees (90°). In a right-angle triangle, we have the following formula commonly known as the Pythagoras Theorem:

$(Adjacent)^2 + (Opposite)^2 = (Hypotenuse)^2$

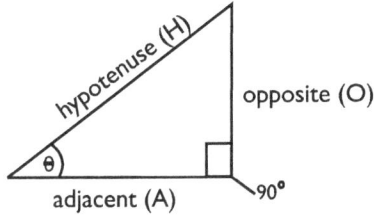

Figure 17.2 – Right-Angle Triangle.

- Every triangle has three interior angles that sum up to 180°.
- The side of the triangle opposite the right angle (90°) is called the hypotenuse, and is the longest side of the triangle.
- The remaining 2 sides, the adjacent and the opposite, are called the legs of the triangle.

17.2.1 Sine (Sin), Cosine (Cos) and Tangent (Tan)

The Sine, Cosine and Tangent functions reveal the shape of the right-angle triangle and are determined as follows:

We use the mnemonic (a pattern that simply helps you to remember something) "SOH CAH TOA" where:

- SOH is used to calculate $Sin(\theta) = Opposite \div Hypotenuse = \frac{O}{H}$.
- CAH is used to determine $Cos(\theta) = Adjacent \div Hypotenuse = \frac{A}{H}$.
- TOA is used to determine $Tan(\theta) = Opposite \div Adjacent = \frac{O}{A}$.

Example

Given the right-angle triangle below, with angle between Adjacent and Hypotenuse equal to θ:

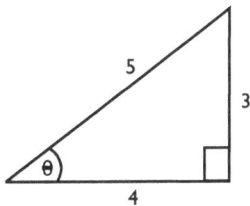

Figure 17.3.

What is Sin (θ)?

Answer

$Sin(\theta) = Opposite \div Hypotenuse = \frac{3}{5} = 0.6$

Application in Real Life

Trigonometry and angles are heavily used in the construction of buildings. For instance, it is used in the determination of angles and slopes of the roof of a building.

17.3 Shapes

17.3.1 Perimeter & Area of a Shape

The perimeter is the distance around a shape.

The area is the measurement of the surface within a shape.

Application in Real Life

Perimeter: In real life, you may apply the calculation of the perimeter to, say, the measurement of the length and width of the boundary of a football field.

Area: You apply the calculation of the area when you may want to calculate the size of, say, a football pitch or the size of a plot of land you have just purchased.

17.3.2 Two-Dimensional Shapes (2-D Shapes)

2-D shapes are shapes that have 2 dimensions only, such as length and width.

Rectangle

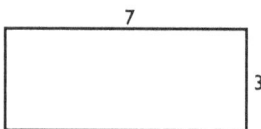

Figure 17.4 – Rectangle with length, 7 and width, 3.

A rectangle is any quadrilateral with opposite sides parallel and equal to each other, and also with four right angles (every angle is 90°); where a quadrilateral is a polygon with four sides and four corners, and a polygon is a closed figure with straight sides.

The perimeter of the rectangle is $7 + 3 + 7 + 3 = 20$ (Here all we have done is, we have added up all the sides of the rectangle)

The Area of the rectangle is $7 \times 3 = 21$.

In general, given a rectangle:

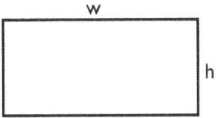

Figure 17.5 – Rectangle with length, w and width, h.

The Perimeter $= w + h + w + h$; where $w =$ length and $h =$ width.

And the Area $= w$ x h.

Pentagon

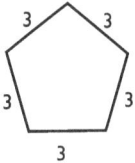

Figure 17.6 – Pentagon with length of each side equal to 3.

A pentagon is any five-sided polygon.

The perimeter of this regular pentagon is $3 + 3 + 3 + 3 + 3 = 15$ (Here all we have done is, we have added up all the sides)

And the area $= \frac{1}{2}$ x Perimeter x apothem

The apothem is a line segment running perpendicular from the centre of the pentagon to the side of the pentagon. Perpendicular here simply means that the apothem is at an angle of 90^0 to the side of the pentagon.

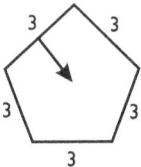

Figure 17.7 – Pentagon showing an apothem.

The pentagon is a five-sided polygon, so there are five apothems.

Triangle

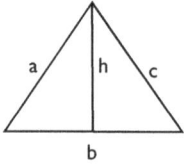

Figure 17.8 – Triangle.

Perimeter = a + b + c (summing the lengths of the sides of the triangle).

Area = (b x h) ÷ 2; where h is the height of the triangle, and b is the length of the base of the triangle.

Square

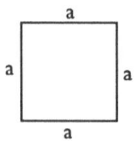

Figure 17.9 – Square.

In a square, all the sides are equal. This differs from a rectangle, where only the opposite sides are equal.

Perimeter = a + a + a + a; where a is the length of each side.

Area = a x a.

Circle

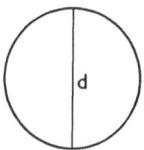

Figure 17.10 – Circle with diameter equal to d.

Perimeter
The perimeter of a circle is called the Circumference.

Circumference (C) = 2 x π x r (where r is the radius and pi (π) represents the ratio of a circle's circumference to its diameter, d, and can also be obtained directly from a scientific calculator by a push of the pi button).

<u>Let's show the steps taken to arrive at the formula for the Circumference.</u>
It is a proven fact that the ratio of the circumference of a circle to its diameter is constant, for any type of circle. You can try this for practice. Therefore, circumference/diameter is denoted π, and is always a constant number roughly equal to 3.142; that is $C \div d = \pi$.

We know that the diameter equals twice the radius, so $d = 2 \times r$.
Hence $C \div (2 \times r) = \pi$ and as such $C = 2 \pi r$.

<u>Area</u>
Area $= \pi \times r \times r = \pi r^2$.

Let's show steps taken to arrive at the formula for the Area.

We divide the circle into 6 pies: 3 black and 3 white.

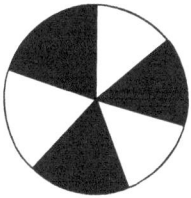

Figure 17.11.

The total length of each black piece in the formed rectangle (see Figure 17.12), using the 6 pieces, is half the circumference ($\frac{1}{2} \times C$). We know $C = 2 \times \pi \times r$, therefore the total length is:

$\frac{1}{2} \times 2 \times \pi \times r = \pi \times r$.

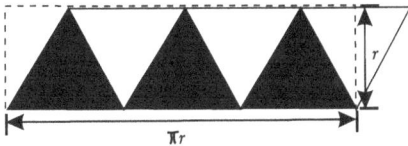

Figure 17.12.

We know the area of rectangle $=$ length x width, so in this case, the length is $\pi \times r$, and the width is r, so the Area $= (\pi \times r) \times r$.

Example 1

A circle has a diameter of 12. Given $\pi = 3.142$:

    a. Calculate its perimeter.

    b. Calculate its area.

Answer

    a. The perimeter of a circle is called the circumference. The circumference C of a circle is equal to $2 \times \pi \times r$. We need to find the radius r. We know $r = \frac{1}{2} \times$ diameter, therefore $r = \frac{1}{2} \times 12 = 6$. Hence $C = 2 \times 3.142 \times 6 = 37.70$.

    b. The area of a circle is given by $\pi \times r \times r = 3.142 \times 6 \times 6 = 113.11$.

Example 2

A room has a length of 100 feet and width of 30 feet. It also has a diagonal measurement of 104 feet.

    a. Calculate the perimeter of the room.

    b. Calculate the area of the room.

Answer

The given measurement of the diagonal is simply additional information and is not necessary for the questions asked.

    a. The room described is rectangular in shape, therefore the perimeter is $100 + 30 + 100 + 30 = 260$ feet.

    b. The area of the room is the length x the width $= 100 \times 30 = 3000$ square feet.

17.3.3 Three-Dimensional Shapes (3-D Shapes)

These are shapes with 3 dimensions, such as height, width and depth.

Cube

A cube has 6 square sides. An example is a cube of sugar. The manufacturers of sugar need the dimensions of area and volume to be able to determine how many cubes to produce in a packet.

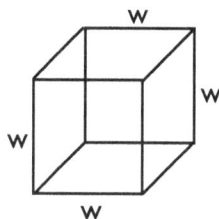

Figure 17.13 – Cube.

Area: The area of the cube is six times the square of the edge length and it is equal to $6 \times w^2 = 6w^2$.

Volume: This is equal to multiplying the edge length three times and it is equal to $w \times w \times w = w^3$.

Perimeter: The cube has 12 sides so this is equal to 12 times $w = 12w$.

Cuboid

A cuboid is similar to a cube, but, rather, has 6 rectangular sides as opposed to 6 square sides. Normally, bars of chocolate are in this shape. Similarly, the dimensions of area and volume are used in the production and pricing the chocolate bar.

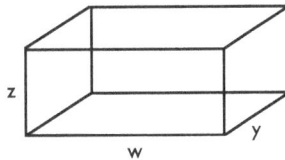

Figure 17.14 – Cuboid.

Area: We add the areas of all six faces to get: (2 x length x width) plus (2 x height x width) plus (2 x length x height).

Perimeter: This is four times the sum of the length, width and height $= 4 \times (w + y + z)$.

Volume: This is the product of the length, width and height $= w \times y \times z$.

Sphere

Figure 17.15 – Sphere.

This is similar to a circle but has 3 axes, namely x-axis, y-axis and z-axis. The shape of a soccer ball is a sphere.

Area: This is given by $4 \times \pi \times r^2$.

Perimeter: Like the Circle, the perimeter is the circumference and it is equal to $2 \times \pi \times r$.

Volume: This is given by $\left(\frac{4}{3}\right) \times \pi \times r^3$.

Cylinder

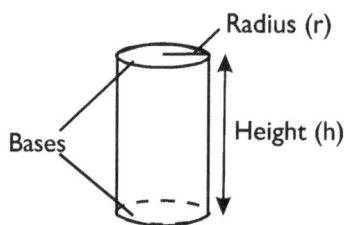

Figure 17.16 – Cylinder.

We use cylinders in our daily lives. The gas cylinder is used in domestic cooking. Also some storage tanks are cylindrical in shape. We need dimensions of area and volume to produce them and price them.

Curved Surface Area: This is the area occupied by the surface of the cylinder, aside of the two bases of the cylinder. Therefore, this is the circumference of a circle times the height of the cylinder which is equal to $(2 \times \pi \times r) \times h = 2\pi rh$.

Total Surface Area: This is equal to the Curved Surface Area plus Area of the two bases $= 2\pi rh + \pi r^2 + \pi r^2 = 2\pi rh + 2\pi r^2$.

Volume: A cylinder can be considered as a stack of circles up to a height, h. So its volume, which includes the hollow space inside the cylinder, is simply the area of a circle times $h = \pi \times r \times r \times h = \pi r^2 h$.

17.4 Geometric Tools

The following are the typical instruments found in a Mathematics box called a Mathematical Set.

The Ruler

A ruler is one of the most widely used Geometric instruments. It is rectangular in shape. Because of how it is shaped, it is used to draw straight lines, and also used in measurement. Remember when we talked about the number line, a ruler is also designed on a number line and this helps us to use it to measure lengths.

Figure 17.17 – Ruler.

The Protractor

This tool is used to measure angles. Its shape is like a half-circle, as you can see from the picture below, and it is used to draw and measure different kinds of angles. It is used to measure any angle within its range. It has two sets of markings; 0 to 180 degrees going from left to right and then right to left. The inner and the outer reading of the protractor are such that the inner and outer readings add up to 180 degrees.

If what we are measuring is on the right side of the protractor, we use the inner readings, and if what we are measuring is on the left side of the protractor, we use the outer readings.

Figure 17.18 – Protractor.

The Compass

This is a two-pronged V-shaped instrument. One prong is stationary, and the other, which holds a pencil, is rotated as we try to trace circles, angles and arcs.

Figure 17.19 – Compass.

The Divider

This looks similar to the Compass, but this has two similar prongs or pointers. They are used to measure lengths between points.

Figure 17.20 – Divider.

Set-squares

Set squares are the triangular pieces of plastic with bits between them taken out. We use them in drawing parallel and perpendicular lines. The two types of set squares are:

- One that has an angle of 45 degrees. This has a right-angle. It is chiefly used to draw vertical lines.
- One that has an angle of 30-60 degrees, which also has a right angle.

Figure 17.21 – Set squares.

Chapter
18

18.0 Approximations

18.1 Rounding numbers to powers of 10

Rounding numbers may mean Rounding up, Rounding down or Rounding off.

18.1.1 Rounding to the nearest 10

Numbers that end in 1 through 4 round to the next lower number that ends in '0'. Numbers that end in a digit of 5 or more should be rounded to the next even ten.

The tens are: 10, 20, 30, 40, 50, 60, 70, 80, 90.

Example 1
Round 14 to the nearest 10.

Answer
14 lies between 10 and 20 (since 14 is more than 10, but less than 20), but 14 is closer to 10 than 20, so the answer is 10.

Example 2
Round 88 to the nearest 10.

Answer
88 lies between 80 and 90 (since 88 is more than 80, but less than 90), but 88 is closer to 90 than 80, so the answer is 90.

18.1.2 Rounding to the nearest 100

To round numbers to the nearest hundred, make the numbers that are in 1 through 49 round to the next lower number that ends in '00'. For example, 424 rounded to the nearest 100 would be 400, This is because the midpoint of 400 and 500 is 450. Since 424 is less than 450, it means it is closer to 400 than 500, so we round it to 400.

Example
Round 475 to the nearest 100.

Answer
475 lies between 400 and 500. Our dilemma is: Do we choose 400 or do we choose 500 as the answer? The midpoint of 400 and 500 is 450. Since 475 is more than 450, it means it is closer to 500 than 400, so we round it to 500.

There may be cases where you are asked to: Round up, Round down or Round off.

Example

We round the number 314765 as follows:

	Round up	Round down	Round off
To the nearest ten	314770	314760	314770
To the nearest hundred	314800	314700	314800
To the nearest thousand	315000	314000	315000

Table 18.1.

Explanation

1. Round up to the nearest ten – First we identify the tens in the number 314765. The tens in the number 314<u>65</u> are as underlined here, i.e. 65. The number 65 lies between 60 and 70, so we round up to 70 to get 3147<u>70.</u>

2. Round up to the nearest hundred – First we identify the hundreds in the number 314765. The hundreds in the number 314<u>765</u> are as underlined here, i.e. 765. The number 765 lies between 700 and 800, so we round up to 800 to get 314<u>800.</u>

3. Round up to the nearest thousand – First we identify the thousands in the number 314765. The thousands in the number 31<u>4765</u> are as underlined here, i.e. 4765. The number 4765 lies between 4000 and 5000, so we round up to 5000 to get 31<u>5000.</u>

4. Round down to the nearest ten – First we identify the tens in the number 314765. The tens in the number 314<u>65</u> are as underlined here, i.e. 65. The number 65 lies between 60 and 70, so we round down to 60 to get 3147<u>60.</u>

5. Round down to the nearest hundred – First we identify the hundreds in the number 314765. The hundreds in the number 314<u>765</u> are as underlined here, i.e. 765. The number 765 lies between 700 and 800, so we round down to 700 to get 314<u>700.</u>

6. Round down to the nearest thousand – First we identify the thousands in the number 314765. The thousands in the number 31<u>4765</u> are as underlined here, i.e. 4765. The number 4765 lies between 4000 and 5000, so we round down to 4000 to get 314<u>000.</u>

Round off simply means to approximate so it could coincidentally become equal to a number rounded up or rounded down. In our case:

7. Round off to the nearest ten – First we identify the tens in the number 314765. The tens in the number 314<u>65</u> are as underlined here, i.e. 65. The number 65 lies between 60 and 70, so we round off to 70 to get 3147<u>70.</u>

8. Round off to the nearest hundred – First we identify the hundreds in the number 314765. The hundreds in the number 314<u>765</u> are as underlined here, i.e. 765. The number 765 lies between 700 and 800, so we round off to 800 to get 314<u>800.</u>

9. Round off to the nearest thousand – First we identify the thousands in the number 314765. The thousands in the number 31<u>4765</u> are as underlined here, i.e. 4765. The number 4765 lies between 4000 and 5000, so we round off to 5000 to get 31<u>5000.</u>

Application in Real Life

In real life a lot of transactions are rounded to whole numbers to make them simpler. If each chocolate sells for $1.99, then the price can be rounded-up to $2 for simpler payments, thus eliminating the need to find coins to give back in change.

18.2 Significant Figures

To be able to determine the quantity of an item, we use significant figures. Generally:

i. Leading zeros are NOT significant. For example, the number 0.0051 has 2 significant numbers.

ii. A zero between two non-zero numbers is significant. For example, the number 505 has 3 significant numbers.

iii. If a number is non-zero, then it is significant. For example, 52.5 has 3 significant numbers.

iv. If a non-leading zero occurs after the decimal, that is to the right of a decimal point, then it is significant. For example, 77.00 has 4 significant numbers.

v. If a zero occurs at the end of a whole number, and there is no decimal point, then it is not significant. For example, 250 has 2 significant numbers.

vi. If a number is exact, then it has an unlimited number of significant numbers. For example, the number 7 can be written as 7.00 (3 significant figures) and is the same as 7.000 (4 significant figures) and is also the same as 7.00000 (6 significant figures) and so on.

18.3 Estimation

We use estimation to guess the value of a calculation. This is usually done mentally to enable us to check or validate the answer to a slightly more complex calculation. Also in most instances, in real life, exact answers are not needed; estimates may be acceptable in instances where someone is simply trying to get an idea of something.

Example
Give an estimate of 73 times 89.

Answer
To do an estimation, we would round the numbers.
We round 73 to one significant figure to get 70, and then round 89 to one significant number to get 90: 70 x 90 = 6,300. This estimate serves as a validation and tells us that the exact answer of 73 times 89 (which is 6,497) is not too far off this estimate of 6,300, depending on the precision we want.

Chapter

19

19.0 Roman Numerals

Number	Roman	Number	Roman	Number	Roman	Number	Roman	Number	Roman
1	I	21	XXI	41	XLI	61	LXI	81	LXXXI
2	II	22	XXII	42	XLII	62	LXII	82	LXXXII
3	III	23	XXIII	43	XLIII	63	LXIII	83	LXXXIII
4	IV	24	XXIV	44	XLIV	64	LXIV	84	LXXXIV
5	V	25	XXV	45	XLV	65	LXV	85	LXXXV
6	VI	26	XXVI	46	XLVI	66	LXVI	86	LXXXVI
7	VII	27	XXVII	47	XLVII	67	LXVII	87	LXXXVII
8	VIII	28	XXVIII	48	XLVIII	68	LXVIII	88	LXXXVIII
9	IX	29	XXIX	49	XLIX	69	LXIX	89	LXXXIX
10	X	30	XXX	50	L	70	LXX	90	XC
11	XI	31	XXXI	51	LI	71	LXXI	91	XCI
12	XII	32	XXXII	52	LII	72	LXXII	92	XCII
13	XIII	33	XXXIII	53	LIII	73	LXXIII	93	XCIII
14	XIV	34	XXXIV	54	LIV	74	LXXIV	94	XCIV
15	XV	35	XXXV	55	LV	75	LXXV	95	XCV
16	XVI	36	XXXVI	56	LVI	76	LXXVI	96	XCVI
17	XVII	37	XXXVII	57	LVII	77	LXXVII	97	XCVII
18	XVIII	38	XXXVIII	58	LVIII	78	LXXVIII	98	XCVIII
19	XIX	39	XXXIX	59	LIX	79	LXXIX	99	XCIX
20	XX	40	XL	60	LX	80	LXXX	100	C
								500	D
								1000	M

The following are general rules applied when writing Roman Numerals:
- The letters I, X and C can only be repeated 3 times in succession.
- We only use I, X and C as subtractive numerals.
- When a lower digit numeral is written to the left of a higher digit numeral, the lower digit is subtracted from the higher digit.
- When a lower digit numeral is written to the right of a higher digit numeral, it is added to the higher digit numeral.

Examples

1. Simplify MML – MCXXXVII, and write your answer in Arabic and Roman numerals.

Answer

First let us find the value of MML in Arabic numerals.

MML = M + M + L = 1000 + 1000 + 50 = 2050.

Next we find the value of MCXXXVII = M + C + X + X + X + V + I + I = 1000 + 100 + 10 + 10 + 10 + 5 + 1+ 1 = 1137.

So our answer in Arabic numerals = 2050 – 1137 = 913.

Therefore, our answer in Roman numerals is:

We break down 913 as:

913 = 900 + 10 + 3 = CMXIII.

2. What is the value of 500 x 2.2 in Roman numerals?

500 x 2.2 = 1100.

We write 1100 as:

1100 = 1000 + 100 = MC.

Application in Real Life

Roman numerals were in use before the current Arabic numerals widely used today. They are still used for things like numbering book chapters and on clock dials these days.

Figure 19.1 – Clock.

Chapter

XX

20.0 Fractional Linear Sequences

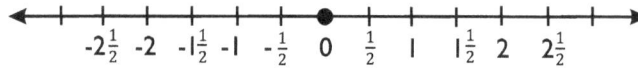

Figure 20.1 – Number line in units of $\frac{1}{2}$.

Example

Given the sequence in Figure 20.1, what is the next number after $2\frac{1}{2}$?

Answer

We subtract 0 from $\frac{1}{2}$ to get $\frac{1}{2}$. Next we subtract $\frac{1}{2}$ from 1 to get $\frac{1}{2}$, and so on. So we see from the sequence, that the numbers are increasing by $\frac{1}{2}$ each time. Hence the next number after $2\frac{1}{2}$ is: $2\frac{1}{2} + \frac{1}{2} = 3$.

Application in Real Life

A calendar year, which is made up of 12 months, is a typical example. At the beginning of January, you are at 0 years. At beginning of February, you have added on $\frac{1}{12}$ years of time. At beginning of March, you have added on another $\frac{1}{12}$ years of time and so on, till you get to the end of December.

Chapter

$$5 \times 2^2 + 1$$

21.0 Numbers

21.1 Prime Numbers

A prime number (or a prime) is a natural number greater than 1, that has no positive divisors other than 1 and itself. Natural numbers are whole numbers that are not negative.

Or

A prime number is a number that can be divided, without a remainder, only by itself and 1.

Examples of prime numbers are: 2, 3, 5, 7, 11, 13, 17, 19, 23, 29, 31.

21.2 Even Numbers

An even number is an integer (not a fraction) that can be divided exactly by 2. The last digit of an even number is always 0, 2, 4, 6 or 8.

Examples of even numbers are: -24, 0, 6, 38.

21.3 Odd Numbers

An odd number is an integer which is not a multiple of two.

Or

An odd number is any integer (not a fraction) that cannot be divided exactly by 2. The last digit of an odd number is 1, 3, 5, 7 or 9.

Examples of odd numbers are: -3, 1, 7, 35.

21.4 Composite Numbers

A composite number is any integer greater than one that is not a prime number.

Examples of composite numbers are: 4, 6, 8, 9, 10, 12, 14, 15.

21.5 Square and Cube Numbers

Given a:
a^2 (pronounced "a squared" or "a raised to the power of 2") = multiplying a by itself 2 times = a x a.

a^3 (pronounced "a cubed" or "a raised to the power of 3") = multiplying a by itself 3 times = a x a x a and so on.

<u>Examples</u>

$10^2 = 10 \times 10 = 100$.

$10^3 = 10 \times 10 \times 10 = 1000$.

Chapter

22

22.0   Converting Decimals to Fractions

Step 1: Write down the decimal divided by 1: $\frac{\text{Decimal}}{1}$ (remember every number can be written as divided by one. For example, the number 0.5 can be written as, $\frac{0.5}{1}$, the number 0.9 can be written as, $\frac{0.9}{1}$, the number 0.33 can be written as, $\frac{0.33}{1}$).

Step 2: Multiply both the top and the bottom by 10 for every number after the decimal point (For example, if there are two numbers after the decimal point, then multiply top and down by 100; if there are three numbers after the decimal point, then multiply top and down by 1000 and so on and forth).

Step 3: Simplify (or reduce) the fraction.

Example

Convert 0.75 to a fraction.

Step 1: Write down 0.75 divided by 1:

$\frac{0.75}{1}$.

Step 2: Multiply both top and bottom by 100 (there are 2 digits after the decimal point in the number 0.75)

$$(0.75 \times 100) \div (1 \times 100) = \frac{75}{100}.$$

Step 3: We simplify $\frac{75}{100}$, by dividing top and bottom by 25 to get:

$\frac{75}{25}$ divided by $\frac{100}{25} = \frac{3}{4}$.

Chapter
23

23.0 Percent

23.1 Writing percentages as a fraction

Percent simply means per 100, since a cent means 100.

We convert a percent to an equivalent fraction with a denominator of 100.

Example

$50\% = 50$ percent $= 50$ per $100 = \frac{50}{100}$.

23.2 Writing percentages as a decimal

Example

50% can be written as 50.0% (Any whole number can be written as such; For example, 11 can be written as 11.0, 6 can be written as 6.0 and so on).

To write 50.0% as a decimal, we move the decimal point two places <u>to the left</u>, to eliminate the % sign.

So we get, after moving 2 places to the left, an answer of 0.5.

23.3 Writing decimals as a percentage

Example

To write 0.5 as a percentage, we move the decimal point two places <u>to the right</u>. Remember 0.5 is the same as 0.50, and can be written as such.

So we get, after moving 2 places to the right, an answer of 50%.

Chapter
6×2^2

24.0 Order of Operations

24.1 BODMAS

The above stands for: (we exclude the "O" in BODMAS above, which just stands for "of" and simply helps to make it a pronounceable word).

1. Brackets
2. Division
3. Multiplication
4. Addition
5. Subtraction

In any given problem, you calculate in the order of 1 first, and then going down in that order, to 5.

<u>Example 1</u>

Solve $(2 \times 1) + 5 - 3$.

<u>Answer</u>
Here, as per our convention, we do the brackets first: $(2 \times 1) = 2$.

Next, we do the addition: $2 + 5 = 7$.

Lastly, we do the subtraction: $7 - 3 = 4$.

So the answer is 4.

<u>Example 2</u>

Solve the problem $2 + 1 \times 3$.

<u>Answer</u>
We do the multiplication first: $1 \times 3 = 3$.

Then, lastly, we do the addition: $2 + 3 = 5$.

So the answer is 5.

Chapter

XXV

25.0 Addition and subtraction of fractions with unlike denominators

The strategy here is to find a common (like) denominator before solving the problem.

Formula

Given: $\frac{a}{b} + \frac{c}{d}$.

The common denominator is b x d, so the formula is:

[a x ((b x d) ÷ b) + c x ((b x d) ÷ d)] ÷ (b x d) = [(a x d) + (c x b)] ÷ (b x d).

Example

Solve: $\frac{1}{4} + \frac{3}{10}$.

Answer
We note here that a =1, b =4, c =3 and d =10, when we plug back into the formula above.

[1 x ((4 x 10) ÷4) + 3 x ((4 x 10) ÷10)] ÷ (4 x 10) = [(1 x 10) + (3 x 4)] ÷ (4 x 10) = $\frac{22}{40}$, which

simplifies to $\frac{11}{20}$, after we divide the numerator and denominator by 2.

Chapter 26

26.0   Division of Fractions

<u>Formula</u>

Given: $\frac{a}{b} \div \frac{c}{d}$.

We transform the above from division to multiplication and solve.

$\frac{a}{b} \div \frac{c}{d}$ becomes $\frac{a}{b} \times \frac{d}{c}$ (note how the $\frac{c}{d}$, has now become $\frac{d}{c}$, with multiplication).

<u>Examples</u>

- $\frac{2}{3} \div \frac{2}{3}$ becomes $\frac{2}{3} \times \frac{3}{2} = \frac{6}{6}$, which when simplified, gives us an answer of 1.

- $\frac{5}{7} \div \frac{9}{16}$ becomes $\frac{5}{7} \times \frac{16}{9} = (5 \times 16) \div (7 \times 9) = \frac{80}{63}$.

Chapter 27

27.0 Least Common Denominator

Example

Given 3 fractions: $\frac{1}{10}, \frac{1}{8}$ and $\frac{1}{6}$, we try to find the Least Common Denominator, i.e. the smallest denominator that these 3 fractions have in common.

The first step is to find the Prime factors of all the denominators.

We take the denominator of, $\frac{1}{10}$, which is 10:
$10 = 2 \times 5.$

Next we take the denominator of, $\frac{1}{8}$, which is 8:
$8 = 2 \times 4 = 2 \times 2 \times 2.$

Then we take the denominator of, $\frac{1}{6}$, which is 6:
$6 = 2 \times 3.$

We then find the prime factors that occur most often and then multiply them together.

2 occurs 3 times when we factor 8, it only occurs once when we factor 10, and occurs once when we factor 6, so:

We pick the three occurrences of 2, from the factorization of 8, to get 2 x 2 x 2.

When we factor 10, we get a 5 which does not occur under factoring 8 or 6 so we pick this single occurrence to get 5.

When we factor 6, we get a 3 which does not occur under factoring 8 or 10 so we pick this single occurrence to get 3.

Therefore, the Least Common Denominator $= 2 \times 2 \times 2 \times 5 \times 3 = 120.$

Application in Real Life

You go out to a grocery store to buy biscuits and sausage rolls for an event. Biscuits come in a pack of 6, and sausage rolls come in a pack of 8. You want to buy the same number of biscuits and sausage rolls, so there is no surplus.

What is the least number of biscuits and sausage rolls you need to buy in order to make sure you are not left with a surplus of either biscuits or sausage rolls?

Task here is to find the Least Common Denominator of 6 and 8.

6 = 2 x 3.

8 = 2 x 4 = 2 x 2 x 2.

Least Common Denominator = 2 x 2 x 2 x 3 = 24

So you need to buy 4 packs (equal to $\frac{24}{6}$) of biscuits, and 3 packs (equal to $\frac{24}{8}$) of sausage rolls.

Chapter

$27 + 1$

28.0   Greatest Common Factor

<u>Example</u>

$48 = 2 \times 2 \times 2 \times 2 \times 3.$
$64 = 2 \times 2 \times 2 \times 2 \times 2 \times 2.$

The greatest common factor is $2 \times 2 \times 2 \times 2 = 16$, since 2 occurs 4 times at most in the factorization of 48 and 64.

<u>Application in Real Life</u>

In a class of pupils, we want to distribute sweets and fruits in such a way that each pupil gets same number of sweets and fruits. If the number of sweets = 120 and number of fruits=100 then to find the maximum number of pupils, we solve as follows:

Here we use Greatest Common Factor of 120 and 100.

$100 = 2 \times 2 \times 5 \times 5.$
$120 = 2 \times 2 \times 2 \times 3 \times 5.$

Greatest Common Factor $= 2 \times 2 \times 5 = 20.$

Chapter

29

29.0 Solving Word Problems

Word problems require you to have a good understanding of language, to help you to structure the problem. They look like they are difficult but in reality all it is asking of you is to be able to extract the key items and show them in numeric form.

Some key terminology, usually found in Word problems, and their mathematical translations are:

Add	Subtract	Multiply	Divide
Combined	Reduced	Product of	Per
Increased	Exclude	Times	Quotient of
Added to	Less	Of	Percent
Increased	Lost		Out of
Bought	Fewer than		Ratio of
Buy	Decreased		Divided by
Together	Take away		Into
Total of	Sold		Shared
Brought in	Sell		
Get	More than		
Received	Spend		
Plus	Minus		
Given	Used		
	Give		
	Gave		

Table 29.1.

Approach

1. Identify the question you need to answer (it is usually the last sentence of the problem, ending with a question mark).
2. Strategize by setting up the Mathematical formula (first by underlining the key points).
3. Solve the problem.
4. Check to see if the result makes sense (For example, if the answer you get for how many hours in one day is 25 hours, then you know your answer likely is wrong, since a standard day cannot possibly have more than 24 hours).

Examples

Question 1
Nii went to the supermarket and bought 20 oranges, but on returning home he had to exclude 7 oranges since they had gone bad. How many oranges did he have left?

Answer
A. Identify the question you need to answer

You solve the question by, first, underlining the key items. It is clear the question we need to solve is "How many oranges did he have left"?

B. Strategize by setting up the Mathematical formula

It says bought 20 oranges so this addition of 20 or +20. (note that +20 is the same as 20. In, mathematics, we normally do not write the + sign in front of a positive number).

It also says exclude 7 oranges so this is subtraction of 7.

C. Solve the problem
Hence the answer is 20 – 7 = 13 oranges left.

D. Check to see if the result makes sense
13 oranges are less than the 20 oranges bought, so the answer makes sense.

Question 2
Nana has lunch 3 times a week at the school cafeteria. She spends 5 dollars out of her 30 dollars weekly allowance every time she has lunch. What fraction of her weekly allowance remains at the end of the week?

Answer
A. Identify the question you need to answer
You solve the question by underlining the key items. It is clear the question we need to solve is "What fraction of her weekly allowance remains at the end of the week?"

B. Strategize by setting up the Mathematical formula

It says 3 times so this is multiplication by 3.

It also says 5 dollars out of 30 dollars so this is: $\frac{5}{30}$.

C. Solve the problem
Hence, fraction spent on lunch weekly is equal to $3 \times (\frac{5}{30}) = \frac{15}{30}$.

So fraction left $= 1 - \frac{15}{30} = \frac{15}{30}$.

D. Check to see if the result makes sense
15 dollars out of 30 dollars left is less than 30 out of 30 dollars she had at the start of the week, so the answer makes sense.

Question 3
Amina <u>received</u> a prize of <u>120 dollars</u> for coming first in her Mathematics exam. She <u>used 20 dollars</u> of the money to buy herself a dress, and <u>shared</u> the remainder equally between her <u>4</u> sisters. <u>How much money did each sister receive?</u>

Answer
A. Identify the question you need to answer
You solve the question by underlining the key items. It is clear the question we need to solve is "How much did each sister receive"?

B. Strategize by setting up the Mathematical formula

It says received 120 dollars so this is addition of 120 or $+120$. (note that $+120$ is the same as 120. In, Mathematics we normally do not write the $+$ sign in front of a positive sign).

It also says used 20 dollars so this is subtraction of 20. And it also says shared equally between her 4 sisters, so this means divide by 4.

C. Solve the problem

Hence the answer is $(120 - 20) \div 4 = 25$ dollars for each sister.

D. Check to see if the result makes sense

The 25 dollars each sister got is less than the remainder of 100 dollars shared between them, so the answer makes sense.

Question 4

Adjo and Tunde were each given a bag containing 6 marbles. On their way back home, Tunde's bag split and lost half of his marbles. Adjo, on her return home gave $\frac{1}{3}$rd of her marbles to her little sister, Abena. How many marbles in total did Adjo and Tunde have left?

Answer

A. Identify the question you need to answer

You solve the question by underlining the key items. It is clear the question we need to solve is "How many marbles in total did Adjo and Tunde have left"?

B. Strategize by setting up the Mathematical formula

It says each given 6 marbles, so we add $6 + 6 = 12$ marbles, to Adjo and Tunde.

It says Tunde lost half of his marbles, so we subtract: $\frac{1}{2} \times 6 = 3$ marbles, from Tunde.

It says Adjo gave away $\frac{1}{3}$rd of her marbles to Abena, so we subtract $\frac{1}{3} \times 6 = 2$ marbles, from Adjo.

C. Solve the problem

Hence the answer is $12 - 3 - 2 = 7$ marbles left.

D. Check to see if the result makes sense

7 marbles are less than the total 12 marbles Adjo and Tunde had in the beginning so the answer makes sense.

Chapter

XXX

30.0 Data

We typically term data as facts and statistics colle cted for analysis or something we can reference on.

Data can either be discrete or continuous.

Discrete data takes integer values only, while continuous data can take any value. For instance, the number of typhoid patients treated by a hospital each year is discrete but your weight recorded over a year is continuous. This is because a patient is a human being and can only take discrete values of 1, 2, 3, 4 etc., but weight changes infinitesimally over a range so it is continuous.

Given the following discrete data of oil being heated:

Time (minutes)	2	4	6	8	10	12	14	16
Temperature (°C)	10	25	35	43	53	65	85	100

Table 30.1.

In isolation, we are given separate data on time and temperature. To make this data meaningful, it is useful to plot the data on a graph so that we can establish a relationship between time and temperature. i.e. how temperature changes over time.

30.1 Line Graph

Title: "*Change of oil temperature with time during heating*"

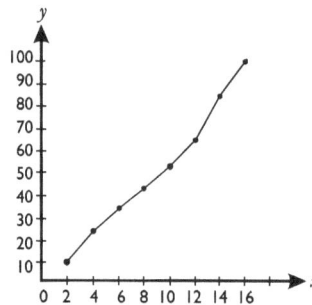

Figure 30.1.

To draw a graph, we have the vertical axis called the y-axis, and the horizontal axis, called the x-axis. The title in this instance is "Change of oil temperature with time during heating" as shown in the graph.
The interval is the space between each value on the scale of the graph.
The time data is on the x -axis and the temperature data is on the y-axis.
We denote the numbers with points in the graph and draw a line through the points.
The shape of the line drawn would tell us what type of relationship exists between time and temperature from the data we have; that is how the temperature of oil increases with time when being heated.

30.2 Bar Graph

Bar graphs differ from line graphs in the way they are constructed. Bar graphs show data of blocks with different lengths and are more useful in showing how often you observe different outcomes than line graphs.

Given the following data:

Name	Age	Favourite subject	Most Disliked subject	Compulsory subject	Shoe Size	Favourite Beverage	Daily Allowance
Zainab	12	English	Mathematics	English	7	Pepsi	5
Kojo	11	Science	Geography	Science	6	Pepsi	10
Tetteh	11	Mathematics	French	Mathematics	8	Fanta	10
Komla	13	Mathematics	History	Mathematics	7	Sprite	10
Tunde	12	Science	Art	Mathematics	8	Coke	7
Jessica	11	Mathematics	Sports	Science	6	Fanta	9

Table 30.2.

We determine the frequency in tabular form, as below. We notice the frequency (number of occurrences) of those aged 11, for example, is 3, since only Kojo, Tetteh and Jessica are aged 11. Similarly, the frequency for those aged 12 is 2, since only Zainab and Tunde are aged 12.

Age	Frequency
11	3
12	2
13	1
Total	6

Table 30.3.

We construct the bar graph, with Frequency on the y-axis, and Age on the x-axis:

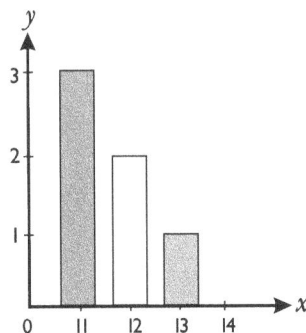

Figure 30.2 – Bar Graph.

We can construct similar bar graphs with the other columns (Favourite subject, Most Disliked subject etc.) as the x-axis, against Frequency on the y-axis, as well.

30.3 Pictogram

Pictograms are graphic images used to represent data. They are similar to bar charts, but use pictures instead of bars.

Solving problems using pictograms

Example

We are given:

= 10, = 5, = 2

Solve:

Solution

Here the answer is $(10 \times 3) \times (5 \times 2) \div (2 \times 2) = 30 \times 10 \div 4 = 75$.

Chapter

31

31.0 Probability

We study probability, to help us to predict possible outcomes in life. It helps us to estimate how much risk is in a course of action we are following, and also helps us to plan for any unforeseen outcomes.

31.1 Experimental Probability

Experimental probability is what **actually** happens when we try out an event. For instance, in a toss of a coin experiment, the coin is physically tossed many times, and as such the probability of a particular side of the coin facing upwards is no longer simply, $\frac{1}{2}$.

31.2 Theoretical Probability

Theoretical probability is what we **expect** to happen when we try out an event. For example, if you have a coin, without tossing it, the probability of one side facing upwards is simply one out of two, $(\frac{1}{2})$, since a coin simply has 2 sides.

Both probabilities are calculated the same way, using the number of possible ways an outcome can occur divided by the total number of outcomes.

31.3 Finding the Experimental Probability

1. We conduct an experiment and we note the number of times the event occurs and the number of times the experiment is performed.
2. We then divide the two numbers to get the experimental probability.

Example
A bag contains 20 red pebbles, 8 blue pebbles and 2 yellow pebbles. Find the experimental probability of getting a blue pebble.

Answer
Take a pebble from the bag and note down its colour, and then return the pebble. Do this about 20 times and count how many times you picked a blue pebble from the bag. We assume this is 13 times. Hence, the experimental probability of getting a blue pebble from the bag is simply, $\frac{13}{20}$.

31.4 Finding the Theoretical Probability

The Theoretical Probability of a given event is the different number of ways the event can occur, divided by the total number of outcomes.

For instance, in the toss of a coin, which has two sides called heads and tails, the theoretical probability is simply, $\frac{1}{2}$.

The experimental probability approaches the theoretical probability of a particular result as the number of trials in an experiment gets bigger

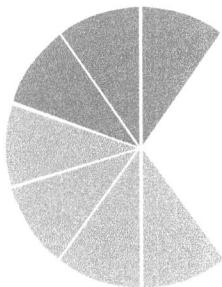

Figure 31.1 – Wheel.

Per Figure 31.1, the total number of pies are 10 and the pies which are Brown are 3. Therefore, the **theoretical probability** of getting a Brown = P(B), which is simply $3 \div 10 = 0.3$.

Now let us see how the experimental probability approaches the theoretical probability, as the number of trials in the experiment increases:

1. If we spin the wheel once, and the outcome is Brown, then $P(B) = 1 \div 1 = 1$. This answer of 1, is much more than the theoretical probability of 0.3, we got before.
2. If we spin the wheel twice, and one outcome is yellow, $P(B) = \frac{1}{2} = 0.5$. We see that this 0.5 probability is nearer to 0.3 now. So we see that the more trials we perform, the closer we get to the theoretical probability of 0.3.

Chapter $(3 \times 10) + (2^1)$

32.0 Positions in space using Cardinal Points

In real life, cardinal points help us to determine where locations are, in relation to other locations.

We draw a diagram showing cardinal points A(-4,4), B(4,4), C(4,-4) and D(-4,-4).

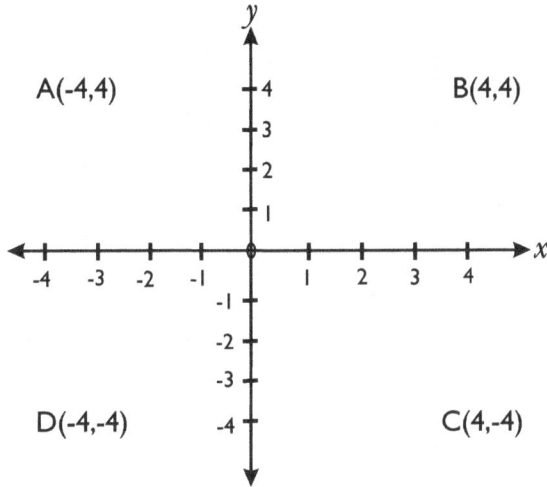

Figure 32.1 – Cartesian plane (X-Y plane) showing the cardinal points A(-4,4), B(4,4), C(4,-4) and D(-4,-4).

Figure 32.1 is labelled as follows:

The vertical axis is called the Y-axis. And the horizontal axis is called X-axis. The center is the number 0. We know all numbers greater than 0 are positive numbers, and all numbers less than 0 are negative numbers.

- The part of the Y-axis north of 0 (above 0) are all positive numbers.
- The part of the Y-axis south of 0 (below 0) are all negative numbers.
- The part of the X-axis east of 0 (to the right of 0) are all positive numbers.
- The part of the X-axis to the west of 0 (to the left of 0) are all negative numbers.

All the cardinal points in the Cartesian plane are denoted by:
1. (-X, Y) – this position implies moving X units west of 0 to get to – X, and then moving Y units north from the position - X, or moving Y units north of 0 to get to Y, and then moving X units west from the position Y.
2. (X, Y) – this position implies moving X units east of 0 to get to X, and then moving Y units north from the position X, or moving Y units north of 0 to get to Y, and then moving X units east from the position Y.
3. (X, - Y) – this position implies moving X units east of 0 to get to X, and then moving Y units south from the position X, or moving Y units south of 0 to get to - Y, and then moving X units east from the position – Y.

4. (-X, - Y) – this position implies moving X units west of 0 to get to – X, and then moving Y units south from the position – X, or moving Y units south of 0 to get to – Y, and then moving X units west from the position – Y.

Examples

For the point A, which is located at (-4, 4), we are saying that A is located at the point where X = -4 on the X-axis meets Y = 4 on the Y-axis in the X-Y plane.

To get to A, we move 4 units to the west of 0 to get to X = -4, and then move 4 units to the north from X = -4.

For the point B, which is located at (4, 4), we are saying that B is located at the point where X = 4 on the X-axis meets Y = 4 on the Y-axis in the X-Y plane.

To get to B, we move 4 units to the east of 0 to get to X = 4, and then move 4 units to the north from X = 4.

For the point C, which is located at (4, -4), we are saying that C is located at the point where X = 4 on the X-axis meets Y = -4 on the Y-axis in the X-Y plane.

To get to C, we move 4 units to the east of 0 to get to X = 4, and then move 4 units to the south from X = 4.

For the point D, which is located at (-4, -4), we are saying that D is located at the point where X = -4 on the X-axis meets Y = -4 on the Y-axis in the X-Y plane.

To get to D, we move 4 units to the west of 0 to get to X = -4, and then move 4 units to the south from X = 4.

End

QUESTION AND ANSWER BANK

QUESTIONS

Question 1
Draw a number line ranging from -10 to 10 in units of 1, and show the position of 8 on this number line.

Question 2
On the number line in question 1, what number do we arrive at when we add -8 to 8?

Question 3
Draw a number line ranging from -10 to 10 in units of 2, and show the position of -6.5 on this number line.

Question 4
On the number line in question 3, what number would you get to if you add 4 to -6.5"?

Question 5
On the number line in question 1, what number would you get to if you add 7 to -2 and then subtract 5?

Question 6
What is the product of 3,000 and 5?

Question 7
Multiply 2,715 by 22.

Question 8
Divide 1,560 by 4 using Short Division.

Question 9
Divide 2,615 by 20 using Long Division.

Question 10
Subtract 3,215 from 4,632.

Question 11
Solve: $\frac{1}{2} - \frac{1}{4}$.

Question 12
Solve: $\frac{1}{2} + \frac{1}{4} + \frac{1}{3}$.

Question 13
Solve: $\frac{1}{3} \times \frac{1}{4}$.

Question 14
Solve: $\frac{1}{2} \div \frac{1}{4}$.

Question 15
Multiply the reciprocal of $\frac{3}{4}$ by $\frac{3}{4}$.

Question 16
Which of the following fractions are equivalent?
$\frac{20}{40}, \frac{1}{3}, \frac{1}{2}, \frac{30}{60}, \frac{2}{3}$.

Question 17
Convert the mixed fraction $2\frac{2}{3}$, into an improper fraction.

Question 18
Which of the following 2 fractions is the largest?
$\frac{2}{3}, \frac{6}{7}$.

Question 19
Put the following fractions in increasing order:
$\frac{1}{3}, \frac{1}{4}, \frac{2}{5}, \frac{3}{7}, \frac{9}{10}$.

Question 20
Multiply $2\frac{2}{3}$ by $\frac{1}{5}$.

Question 21
Solve for x, in the equation:
$6x + 5 = 17$.

Question 22
Solve for x in the equation:
$6x - 5 = 17 - 7y$.

Question 23
Solve for x in the equation:
$\frac{3x+6y}{y} + 5 = 17$.

Question 24
Given y = -6, and $6y - 6 = 17x + c$.
What is x ?

Question 25
Given $\frac{3x}{3} = \frac{(7+y)}{3}$, solve for x.

Question 26
Given a < b < c < d, which one is the greatest?

Question 27
Given a < b < c < d, which one is the least?

Question 28
Solve for x in:
7x – 5 = 22.

Question 29
Solve for x in:

7x – 5 = 22y.

Question 30
9 < x < 11. If x is a whole number, what is x?

For questions 31 to 34, given the number 3,653,036.22:

Question 31
Which digit is in the millions position?

Question 32
Which digit is in the one hundredth position?

Question 33
Which digit is in the billions position?

Question 34
Divide the number in the ten thousandth position by 100.

Question 35
Write $\frac{2}{5}$ in decimal form.

Question 36
Calculate $\frac{2}{3}$ to three decimal places.

Question 37
Multiply 0.25 by 0.35.

Question 38
Divide $\frac{2}{3}$ by 0.25.

Question 39
Which of the following are valid if "a" and "b" are each greater than 1?
- i.   a + b = b – a.
- ii.   a x c = c x a.

iii. $\frac{a}{c} = \frac{c}{a}$.

iv. $a + b = b + a$.

Question 40

Is the following valid?

$(a - c) - b = a - (c + b)$.

Question 41

Which of the following are valid if "a" and "b" are each greater than 1?

    i. $a(b + c) = ab + ac$.

    ii. $a(b + c) = ab + c$.

Question 42

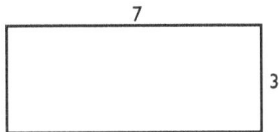

Given the rectangle above, what is the area?

Question 43

What is the perimeter of the rectangle in question 42?

Question 44

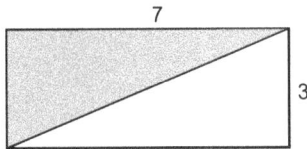

Given the rectangle above, what is the area of the shaded portion?

Question 45

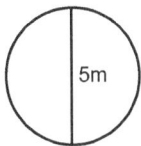

Given the circle above, with diameter = 5m, what is the area?

Question 46
What is the perimeter of the circle in question 45?

Question 47
What is a polygon?

Question 48
Describe a quadrilateral.

Question 49

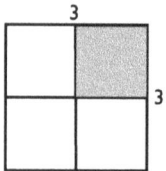

You are given a square with length of each of the sides of 3. Find the area of the shaded portion, if each quadrant within the square is of the same size.

Question 50
Explain Pi.

Question 51
Express 213.33 centimeters in meters.

Question 52
What is 500.23 grams in kilograms?

Question 53
What is the sum of 170 decimeters and 777.7 centimeters, in meters?

Question 54
A rectangle measures:

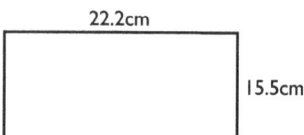

Calculate the area in meters.

Question 55
What is the y-axis?

Question 56
What is the x-axis?

Question 57
Given the following Cartesian plane, is the position A(6,6) identifiable?

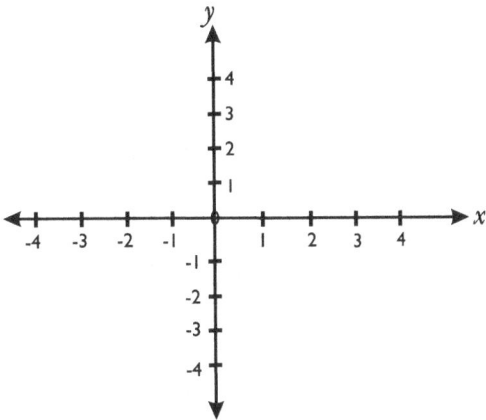

Question 58
In the Cartesian plane in question 57, is the position B(-2,-8) identifiable?

Question 59
Complete the square in the diagram below. i.e. what is the coordinate that completes the square, with vertices marked "x" ?

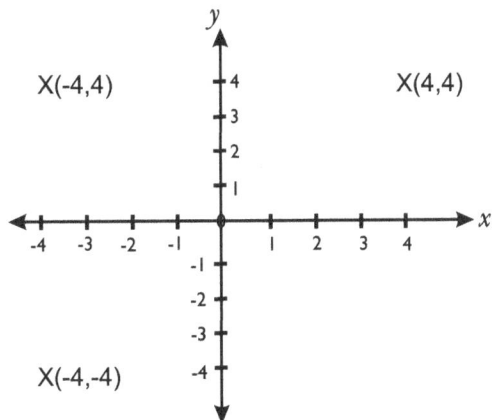

Question 60
What is MCMXXII in Arabic numerals?

Question 61
Subtract 1,922 from MCMXXII and give your answer in Arabic numerals.

Question 62
Write XIX in Arabic numerals.

Question 63
Calculate (LIX)² in Arabic numerals.

Question 64
Which one is the greater of IX and XI?

Question 65
Subtract L from MCMXXIV and give your answer in Roman numerals.

Question 66
A train heading to Kano from Accra had 477 passengers. 300 passengers left the train at Lagos and 150 passengers boarded the train then. How many passengers in total made it to Lagos from Accra?

Question 67
Jim received 30 apples as a present from his grandmother. 30% of the apples were red, 20% of the apples had gone bad on arrival. How many apples had gone bad?

Question 68
For her birthday, Ama received a purse worth 50 dollars. In the purse was also a 50 dollar note which she spent on lunch and dessert. How much money in total did she receive?

Question 69
10 lights bulbs are delivered daily to a depot. 25% of the bulbs are red coloured bulbs and 15% of the remainder are yellow coloured bulbs. The rest of the bulbs are non-coloured. How many non-coloured bulbs are delivered to the depot over 3 days? (Round your answer to the nearest integer).

Question 70
A large piece of land, rectangular in shape with length 100m and width 70m was given to Tom. Tom decided to give 30% of the land to his sister Esi. What is the area of the remaining land?

Question 71
What is a prime number?

Question 72
What is the largest prime number less than 1,000?

Question 73
What is an even number?

Question 74
What is the smallest even number?

Question 75
What is an odd number?

Question 76
What is the greatest odd number less than 500?

Question 77
What is the least common multiple of the following numbers: 5, 6, 7?

Question 78
What is the greatest common factor of the following numbers: 10, 11, 12?

Question 79
Divide 25.5 by 5.5, and express your answer in fraction.

Question 80
Given $x + x = 50$, what is x?

ANSWERS

Question 1

Question 2
If we add -8 to 8, we are counting eight units to the left and we arrive at 0.

Question 3

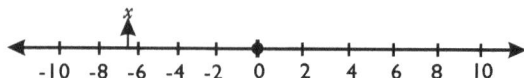

Question 4
Since the units are 2 each, we move $\frac{4}{2}$ units to the right from -6.5. One unit takes us to -4.5 and 2 units takes us to -2.5.

Question 5
0.

Question 6
As you may recall from Chapter 8, Product means Multiplication. Therefore, we multiply 3,000 by 5 to get 15,000.

Question 7
59,730.

Question 8
390.

Question 9
$130\frac{3}{4}$.

Question 10
1,417.

Question 11
$\frac{1}{4}$.

Question 12
$\frac{13}{12}$.

Question 13

$\frac{1}{12}$.

Question 14

2.

Question 15

1.

Question 16

$\frac{20}{40}$, $\frac{1}{2}$ and $\frac{30}{60}$ are equivalent.

Question 17

$\frac{8}{3}$.

Question 18

$\frac{6}{7}$ is higher.

Question 19

In increasing order: $\frac{1}{4}$, $\frac{1}{3}$, $\frac{2}{5}$, $\frac{3}{7}$, $\frac{9}{10}$.

Question 20

$\frac{8}{15}$.

Question 21

$X = 2$.

Question 22

$X = \frac{(22-7y)}{6}$.

Question 23

$X = 2y$.

Question 24

$X = \frac{(-42-c)}{17}$.

Question 25

$X = \frac{(7+y)}{3}$.

Question 26
d.

Question 27
a.

Question 28
$X = \frac{27}{7}$.

Question 29
$X = \frac{(22y + 5)}{7}$.

Question 30
10.

Question 31
3.

Question 32
2.

Question 33
None.

Question 34
0.

Question 35
0.4.

Question 36
0.667.

Question 37
$\frac{7}{80}$.

Question 38
$\frac{8}{3}$.

Question 39
ii and iv.

Question 40
Yes.

Question 41

i.

Question 42

21.

Question 43

20.

Question 44

10.5.

Question 45

19.63m².

Question 46

15.71.

Question 47

Bookwork.

Question 48

Bookwork

Question 49

2.25.

Question 50

Bookwork.

Question 51

2.1333m.

Question 52

0.5 kg.

Question 53

24.77m.

Question 54

0.0344m².

Question 55

It is the vertical axis on the Cartesian Coordinate plane.

Question 56

It is the horizontal axis on the Cartesian Coordinate plane.

Question 57
No.

Question 58
No.

Question 59
(4, -4).

Question 60
1922.

Question 61
0.

Question 62
19.

Question 63
3,481.

Question 64
XI.

Question 65
MDCCCLXXIV.

Question 66
477.

Question 67
6 apples.

Question 68
50 dollars.

Question 69
19 (rounded to the nearest integer).

Question 70
4,900 square meters.

Question 71
Bookwork.

Question 72
997.

Question 73
Bookwork.

Question 74
2.

Question 75
Bookwork.

Question 76
499.

Question 77
210.

Question 78
1.

Question 79
$\frac{51}{11}$.

Question 80
25.

Notes – Chapter 8

Notes – Chapter 24